Introduction to the study of biological membranes

MARCELINO CEREIJIDO

and

CATALINA A. ROTUNNO

GORDON AND BREACH SCIENCE PUBLISHERS
New York · London · Paris

150 Fifth Avenue, New York, N.Y. 10011

Editorial office for the United Kingdom:
Gordon and Breach, Science Publishers Ltd.
12 Bloomsbury Way
London W.C. 1

Editorial office for France:
Gordon and Breach
7-9 rue Emile Dubois
Paris 14e

As well as belonging to the University of Buenos Aires, the authors are also Career Investigators for the Consejo Nacional de Investigaciones Científicas de la Argentina (National Research Council).

FIRST PRINTING 1970
SECOND PRINTING 1971

Library of Congress catalog card number: 71–101309. Standard book number 677 02410 X.

Printed in the United States of America

Introduction to the study of biological membranes

To our friend
Prof. Dr. Hersch M. Gerschenfeld

Contents

Introduction

According to the second law of thermodynamics the distribution of matter and energy in the universe tends to become less and less orderly and more and more of the sort that would result from the operation of the law of chance. From this point of view, even the simplest organism is an almost incredibly improbable accumulation of matter

M. JACOBS (1935)

IMAGINE A unicellular organism suspended in an aqueous environment. A sodium ion penetrating into the cell becomes a part of one of the most complex systems in the universe: a living organism. When the sodium leaves the cell it dilutes in one of the simplest systems: water. The tremendous transition between the two worlds takes place at the level of a delicate layer one millionth of a centimetre in thickness: the cell membrane. The "continued existence of the cell is dependent on the ability of this membrane to permit passage of some substances and prevent that of others" (Davson and Danielli, 1943). How does the cell membrane recognize sodium, potasium water, glucose, a vitamin, a hormone among thousands of substances? What mechanisms does it use to make them penetrate or leave the cell?

Living systems are constituted of compartments separated from each other and from the external environment by membranes and other barriers. A bacterium, the plasma, a neuron, the whole brain, a mitochondrion are examples of such compartments. Each one achieves and maintains a characteristic steady state composition, usually completely different from the composition of its surroundings, thanks to a delicate balance between the influx and outflux of matter and energy through their membranes.

The existence of compartments limited by membranes may also determine the characteristics of a biochemical reaction. If, in a homogenized tissue, substances *A* and *B* react to form *C*, it does not necessarily mean that the reaction is physiologically important. It could happen that the substances were in different compartments and could not pass through the membranes.

Enzyme 2 of figure I-1 binds substance S_b at site B and converts it into S_c, but is inhibited when S_b attaches to site A. If this reaction were studied in a homogenized tissue where S_b has access to site A, Enzyme 2 may not work and the whole reaction may be different. It might also happen that *in vivo* the reaction $S_a \rightarrow S_b \rightarrow S_c$ is rate limited by the diffusion of S_a through the

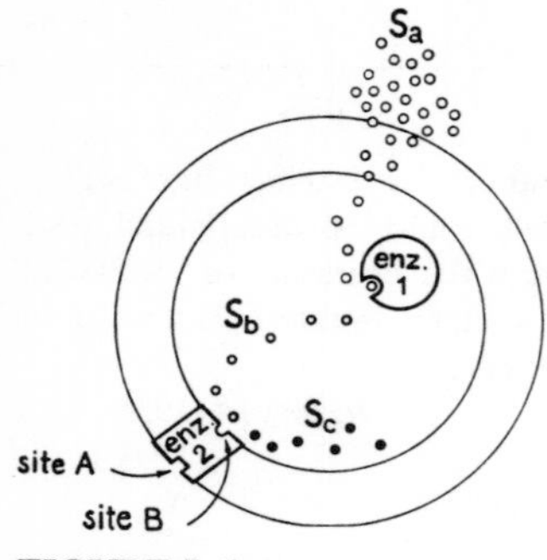

FIGURE I–1

membrane while, if we purified the enzymes, we might find that the reaction is limited by the conversion of S_b into S_c. Many hormones govern a metabolic reaction by regulating the permeability of a membrane to a given substrate. As Siekevitz pointed out "... we are entering a new phase of biochemistry in which extracted purified enzymes are going to be added back to the cell structure from which they were initially obtained".

Membranes act as selective barriers and it was this property which led to their discovery. However their biological role is not reduced to separate compartments. The abundance and arrangement of membranes inside chloroplasts, mitochondria, retinal rods, or in the cytoplasm in contact with the ribosomes (figure I-2) suggest that membranes participate in the most fundamental biological processes. Membranes have an elaborated organization in dynamic equilibrium with the environment, and are specially suited to receive, store, convert and transfer information. It looks as if, in order to perform complicated chemical reactions, nature designed molecular circuits and mounted the participating molecules on membranes.

Knowledge of what a membrane is, and how it works helps to understand why a substance is absorbed in the intestine, why a bacterium metabolizes a given amino acid, how the kidneys produce urine, how ATP is synthetized in the mitochondrion, why the root of a plant absorbs a given salt from the ground, and why a neuron receives and transmits electrical signals.

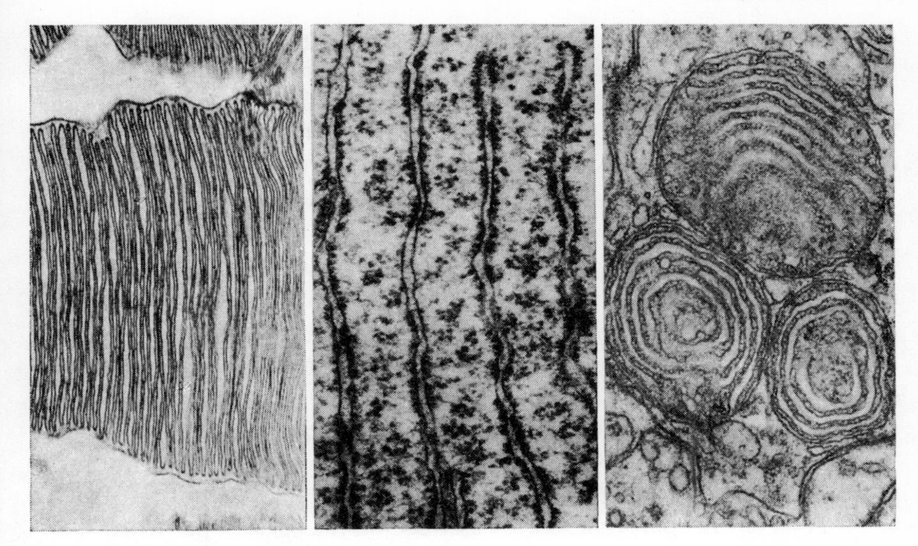

FIGURE I–2 Circumstances in which membranes do not seem to be acting merely as barriers. *Left:* retinal rod of rat. (× 30.000). *Center:* endoplasmic reticulum of a neuron of monkey (× 60.000). *Right:* mitochondria of the pineal gland of a rat (× 60.000). Reproduced by courtesy of Drs. A. Pellegrino de Iraldi and J. Pecci Saavedra.

Having emphasized the role of membranes as barriers we may now ask: is the cytoplasm in a state of free solution and the movement and distribution of substances solely determined by the properties of the cell membrane? The answer is *no*. As our knowledge of the structure of water, and of the relationship of water, ions and macromolecules inside the cell improves, it becomes evident that the cell is not just a droplet of well stirred suspension, but has a high degree of organization. The accumulation or exclusion of many substances from the cell, that were generally thought to be entirely due to the operation of membrane mechanisms, may in fact reflect their specific binding to macromolecules of the protoplasm or their poor solubility in a highly polarized cellular water. Unfortunately, the view that the membrane is an universal barrier which determines the rate of selective entry and exit of water and solutes into and out of the cell as well as their concentration levels, and the opposite view that the steady state levels of solutes and water reflect the properties of the whole protoplasm, have been developed in parallel fashion, almost without dialogue. We feel that if we forced a merge and tried to describe a unified picture, we would have distorted the present status of the problem. Our plan is to present each subject as described by the workers in the specific field: we describe the Na-pump as scientists working with ATPase view it, and Ling's hypothesis as if the Na-pump did not exist. We hope, though, that we will give the reader enough flexibility as to accept opinions and criticisms from both sides.

The subject is also a no-man's-land between enzymology, electronics, crystallography, electron microscopy, thermodynamics, cell physiology, etc. Contributions come from the study of subjects as diverse—and apparently not related—as physical chemistry of glasses and organic chemistry of macrocyclic antibiotics, phase transitions in soaps and electrical properties of ice, hydrolysis of ATP and intercellular communication. Also the differences in information, assumptions, technology, and language used in each of these fields, make difficult the understanding among workers.

The emphasis of the book is on general phenomena and modern views, rather than on their application to specified physiological problems. Inevitably, however, the phenomena are described on the basis of particular biological preparations in which the studies were orginally observed or are more clearly expressed. Its aim is to take a student, or a non-specialist, without previous knowledge of membranes and enable him to read original papers with some understanding. The bibliography is not meant to endorse what the text says, but to provide a reading list to complete the information.

While we must alone bear responsibility for the views expressed in this book, it gives us much pleasure to express our gratitude to Drs. G. Eisenman, G. N. Ling, V. Luzzatti, and W. F. Widdas for their suggestions and constructive criticism of the chapters in which their leading ideas are described, and to Drs. A. S. Frumento, P. Garrahan, H. M. Gerschenfeld, M. I. Pouchan, A. Rega, I. Reisin, E. Rivas, and F. A. Vilallonga for reading the manuscript, making suggestions for improving obscure passages, furnishing examples, checking equations, and other help. Our work in the field was made possible through the generous aid of the Consejo Nacional de Investigaciones Científicas of Argentine, The Public Health Service and The Population Council of U.S.A., and Gerardo Ramón Laboratories.

List of symbols

Unless otherwise stated, the meaning of the symbols used is the following

Symbol	Meaning
Å	ångström (10^{-8} cm)
C_i	concentration of i; sometimes it is also noted as i to avoid crowding of subindices, like K_{∞}^{cell} (*ie*, the concentration of potassium in the cell in steady state)
D	diffusion coefficient
F	the Faraday (96.500 coulomb/mole)
G	Gibbs' free energy
I	electric current
J_{kj}^{i}	flux of substance i from compartment k to compartment j.
k	Boltzmann's constant
k_{ij}	rate coefficient. Fraction of S_i transferred to compartment j.
L_{ij}	phenomenological coefficient relating the ith flux to the jth force
n_i	number of moles of component i
$\mathscr{P}$	permeability
P	pressure
P_i	amount of tracer in compartment i
p_i	concentration of tracer in compartment i
p_i^*	specific activity of tracer in compartment i
R	gas constant
S	entropy
S_i	amount of substance in compartment i
T	absolute temperature
t	time
U	internal energy
u_i	mobility of ion i ($\text{cm}^2\ \text{sec}^{-1}\ \text{volt}^{-1}$)
v	volume
x	distance (usually the thickness of the membrane)
X_i	thermodynamic force acting on component i

1 Cereijido/Rotumno (0241)

z_i	valence of component *i*
μ_i	chemical potential of component *i*
$\bar{\mu}_i$	electrochemical potential of component *i*
μl	microliter
μmole	micromole
μM	micromolar
ψ	local electrical potential
ω_i	mobility of component *i* (cm mol sec^{-1} dyn^{-1})
ADP	adenosinediphosphate
ATP	adenosinetriphosphate
ATPase	adenosinetriphosphatase
Sub-index	0 initial ∞ steady state 1, 2, 3... compartments 1, 2, 3...

1

The structure of the membrane

THE CELL membrane was born as a functional necessity in the second half of the last century. That was a period in which the attention was focused in whether a substance penetrated or not into the cell and, in case it did not, what prevented it from entering. The observation that cells swell in hypotonic solutions and shrink in hypertonic ones led to the comparison of the cell with an osmometer. The analogy required a membrane. Pioneers in this field were the studies of Nagelli and Cramer (1855) Pfeffer (1877) DeVries (1884, 1855) Hedin (1897, 1898) Overton (1899) and others. Some of them though, accepted the existence of a cell membrane reluctantly or did not accept it at all. The major objection was that nobody had seen it. The not acceptancy of the membrane on this basis does not surprise if one considers that the membrane is through to be some 100 Å thick and, therefore, it takes an electron microscope to "see" the membrane.

Overton (1899) pointed out that the cell was more permeable to substances which dissolved easily in lipids than to substances that have poor lipid solubility. Consider for instance figure 1-1, which is a famous figure of Collander (1949). It shows that the penetration of a substance into the cell (expressed as permeability times the square root of the molecular weight) is roughly proportional to its olive oil-water partition coefficient. Therefore, in a first approximation the cell membrane was considered to be a lipid film that substances have to traverse in order to get into the cell. A few decades later Gorter and Grendel (1925) extracted the lipids from the erythrocytes, spread them as a monomolecular film on a water subphase and observed that the ratio of film area to erythrocyte area from which the lipids were extracted was 2:1. It was then considered that the erythrocytes have sufficient lipid to form two monomolecular layers (a bilayer) around them. Since most membrane lipids have a hydrophilic and a hydrophobic portion, the most obvious model for a lipid membrane is the one depicted in figure 1-2. This arrangement configurates a membrane of thickness equal to two lipid molecules, *i.e.* some 50 to 60 Å.

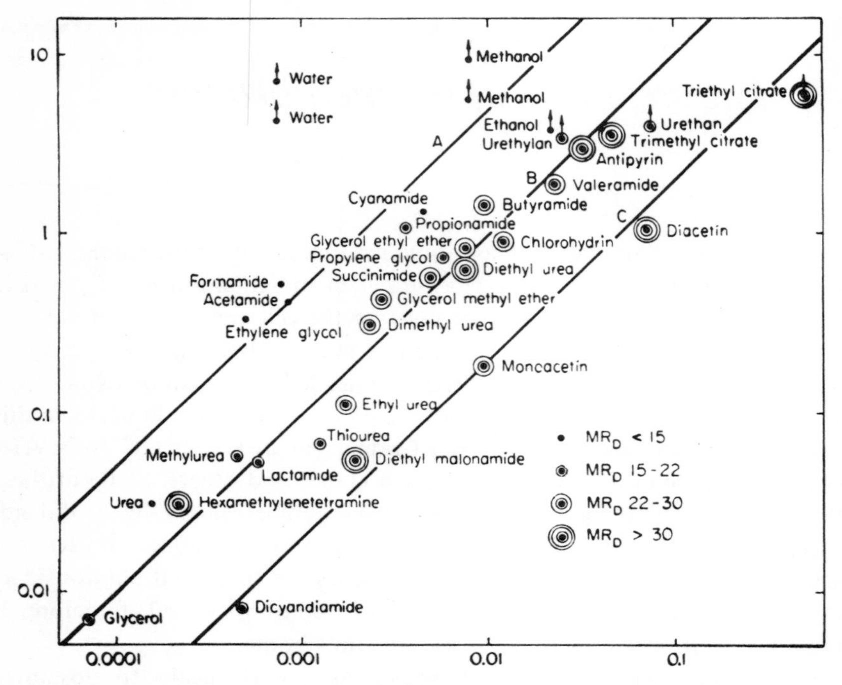

FIGURE 1–1 Permeability of non-electrolytes across the membrane of *Chara vs* their oil water partition ration. Permeability is multiplied by the square root of the molecular weight. MR_D is a function of the molecular size (Taken with kind permission from R. Collander, 1949).

Fricke (1925 and 1933, see also Cole, 1950) observed that when a low frequency current is passed through a suspension of cells in a saline solution the cells behave as non-conducting bodies, but as the frequency is increased, the impedance decreases. To explain these results he adopted a model in which the cells were surrounded by a thin layer of an insulating material. Using a modification of Maxwell's equation and assuming that the insulating material had a dielectric constant of 3 (an acceptable value for lipids) he found that the layer had to be about 50 Å thick to account for the capacity of the suspension, a result that agrees remarkably well with the model of figure 1-2.

By that time Harvey (1931) studied the surface tension of cells subject to a centrifugal force. Under the influence of this force the light and the heavy components of the cell separated producing an elongation of the cell. As the speed of the centrifuge was increased the separation was more pronounced until a centrifugal force was reached (0.2 dyne $\times$ cm^{-1}) that broke

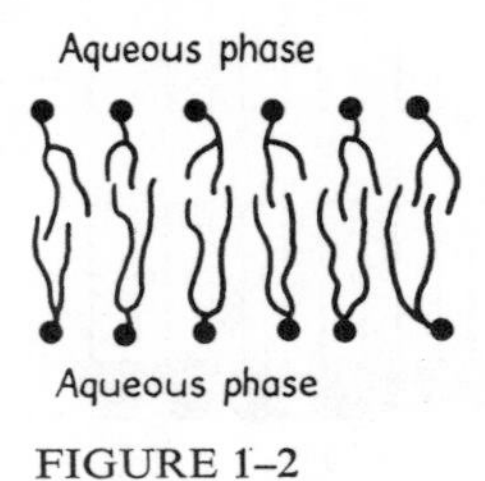

FIGURE 1–2

the cells into two halves. Since the cell membrane has to be broken in order to separate the two pieces of cell, the force necessary to split the cell is a measure of the surface tension of the membrane. This surface tension proved to be too low as the one expected for a simple oil–water interface (some 35 dynes $\times$ cm^{-1}). It was later demonstrated that the tension was lowered because of the adsorption of proteins to the lipid bilayer as depicted in figure 1-3 (see Danielli and Harvey, 1935; Harvey and Danielli, 1938; Askew and Danielli, 1940 and Cole, 1932).

An important support for this model was provided by the works of Schmitt, Bear and Ponder (1936) who studied the refrigence of cell membranes before and after the extraction of lipids and concluded that the lipids are oriented in a perpendicular position to both the surface of cell and the protein component of the membrane.

After this glimpse to almost a century of cell membrane history, we have a general model where to base our discussion. We will now abandon the historical approach in favour of a more analytical one. Our plan for the study of the membrane structure is the following; we will study first the chemical composition of the membrane. Then we will review a series of experimental preparations that are commonly used in the study of the physico-chemistry of cell membranes and of the interaction occurring in lipids, water-lipid, and water-lipid-protein systems. At that point we will have enough information as to reconsider Davson-Danielli's model (the one depicted in figure 1-3) and to discuss further improvements or other alternatives. Finally we will discuss briefly some recent information on the dynamical aspects of the membrane architecture.

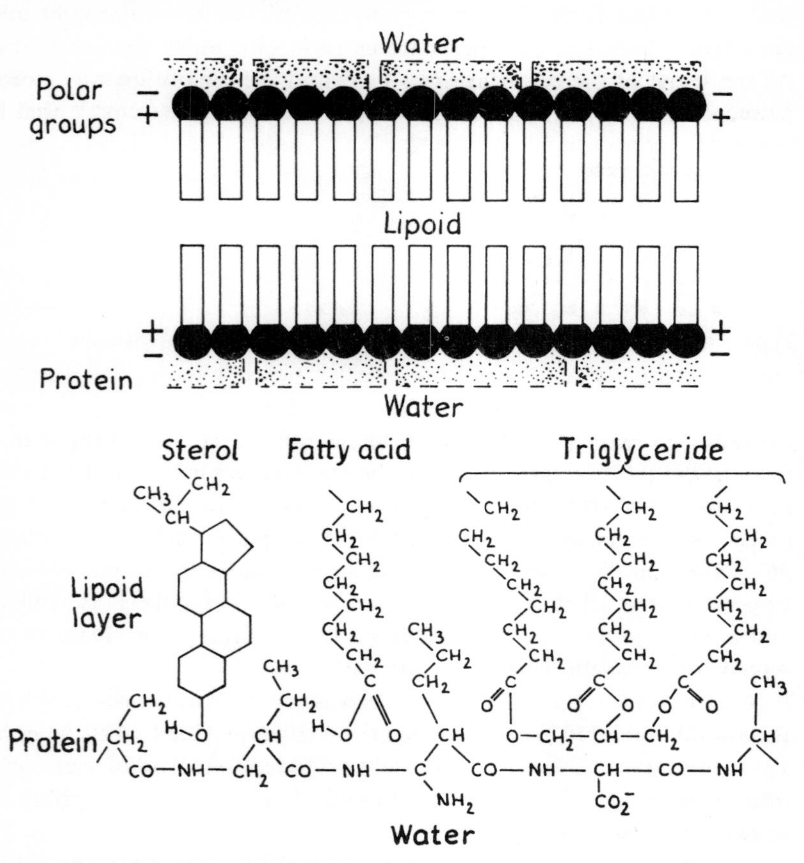

FIGURE 1–3 General pattern and chemical detail of the Davson and Danielli's model.

1.1 Chemical composition of the membrane

In order to analyze the chemical composition of the membrane one has first to separate it from the rest of the cell. The cells are usually broken by sound, hypotonic shock or mechanical homogenizers. The disrupted plasma membrane rearranges into fragments in the form of small vesicles.

They are then separated from the rest of the cellular components by flotation, centrifugation in density gradients or extraction with buffer solutions. Some preparations offer special advantages in this respect. In the case of the red cells, for instance, they can be treated with hypotonic solutions which swell the cells, loosen the membrane framework, and allow the cell content to escape leaving an empty *ghost* constituted by little more than the cell membrane. In the case of the nerve myelin, the sheaths of membrane account for about 80 per cent of the total dry mass (Finean, 1960).

In extracting the membrane several questions arise. How do we know, for instance, that the extract contains the whole membrane? Did some membrane components go into the discarded cellular fractions? Does the extract contain peripheric cell membrane only? At least in the Ehrlich ascites carcinoma cells, the peripheric cell membrane fragments into diverse vesical types, each bearing a different fraction of the intact cell surface (which, by the way, tends to indicate that the cell surface of these cells is organized as a mosaic of large and functionally discrete macromolecular assemblies [Wallach, 1965]). The different fractions containing surface membrane also contain other cell constituents, mainly intracellular cell membranes, that have to be excluded by further purifications.

To solve these problems the purification is controlled by electron microscopy (Emmelot and Bos, 1962; De Robertis, 1967), by following the purification of surface antigens (Herzenberg and Herzenberg, 1961; Wallach and Kamat, 1964), by following the purification of enzymes known to be attached to the membrane (McCollester and Randle, 1961; Wallach and Kamat, 1964) or by tracing "membrane markers", *i.e.* substances that bind specifically to the cell membrane and do not penetrate the cell (Maddy, 1964; Marinetti and Gray, 1967). For a discussion of the purification-recognition problem see: Wallach, 1965; Maddy, 1966a and b. A corollary of this situation is that our knowledge of the chemical composition of the membrane depends, to certain extent, on when the biochemist stops the purification and starts the analysis.

The three major components of the membranes are: proteins, lipids, and water. Although the different membranes have the same general pattern, they vary widely in the proportion in which proteins and lipids enter into their composition. For references on the chemical composition of the membrane see: Kono and Colowick (1961); Razin, Argaman and Avigan (1963); Autilio, Norton and Terry (1964); Salton and Freer (1965); Maddy (1966a).

1.1.1 The protein component

Proteins may constitute up to 80% of the cell membrane (*e.g.*: the red cell membrane) or as little as 18% (*e.g.*: nerve myelin). Variations are even wider in bacteria. One of the membranes which can be better purified is the erythrocyte's. Treated with a variety of organic solvent and buffer solutions they yield proteinaceous extracts: *stromatin* (Jorpes, 1932), *elenin* (Moskowitz and Calvin, 1952) etc. Other workers have resorted also to the action of urea, dialysis, detergents, etc. (Maddy, 1964, 1966b). Even when, in most cases, it was possible to obtain extracts with enzymatic activity (McCollester and Randle, 1961); Wallach and Kamat, 1964; Emmelot, Bos, Benedetti and Rumke, 1964) or with contractile activity (Onishi, 1962), the general observation is that purification reaches a point where the proteins lose their specific properties. This perhaps is due to the fact that enzymes attached to a support like the membrane, have a high degree of organization and special properties that are lost once the membrane is disjoined.

1.1.2 The lipid component

The lipid component also varies widely from membrane to membrane, constituting up to 80% of the myelin, but only 15% of the skeletal muscle membrane. Not only the total lipid component varies, but also the chemical nature of the lipids. Thus in Escherichia coli, Agrobacterium tumefaciens and Azobacter agilis, phospholipids constitute 100% of the lipid phase, but in the myelin membrane phospholipids are a mere 38%. Cholesterol constitutes some 25% of the lipids of the erythrocyte membrane, but it does not exist in Escherichia coli, Agrobacterium tumefaciens and Azobacter agilis. Triglicerides and other lipids, which are absent in these bacteria, constitute a 38% of the lipids of the myelin (Kaneshiro and Marr, 1962; Ways and Hanahan, 1964; O'Brien and Sampson, 1965). Table I and II illustrate the nature of the polar groups and the length and saturation of the hydrocarbon chains commonly found in membranes. They vary widely from membrane to membrane. As we shall see later, the charges of the polar groups and the length and saturation of the chains bear profound influence in the permeability and other properties of the membranes.

1.1.3 The water component

In trying to detect the tiniest bit of a substance suspected to be responsible of a given biological reaction, a biologist is likely to resort, if needed, to the fanciest chemical and physical technology available. All the while he may have

TABLE I Major phosphatide components (ranges, in percent, in erythrocyte membranes)[a]

Lecithin	1 (sheep)	56 (rat)
Cephalin	18 (rat)	46 (pig)
Sphingomyelin	14 (pig)	63 (sheep)

The major fatty acids are: *Palmitic* (16 carbons), *Stearic* (18 carbons), *Oleic* (18 carbons, 1 double bond in C_9), *Linoleic* (18 carbons, 2 double bonds: in C_9 and C_{12}) *Linolenic* (18 carbons, three double bonds in C_9, C_{12} and C_{15}), *Arachidonic* (20 carbons 4 double bonds: in C_5, C_8, C_{11} and C_{14})

[a] data from: Dawson, Hemington and Lindsay (1960), de Gier and Van Deenen (1961) Dodge and Phillips (1967), Cornwell, Heikila, Bar and Biagi (1967)

TABLE II Polar head groups of lipid molecules in membranes (% of total lipid in membranes)[a]

		Human myelin	Human erythrocyte
Uncharged	non-polar (cholesterol, triglycerides)	40.1	43.9
	polyhydroxyl	16.0	6.0
Zwitterionic		15.4	24.4
Weak acids (carboxylic, and phosphoril-ethanolamine)		18.4	18.7
Strong acids		9.1	7.0

[a] data from: O'Brien (1967)

right in front of him several moles of water with no other assigned role than to serve as a stage for the biological reaction. Although life does not occur in the absence of water, most reactions and physiological events are explained on the basis of enzymes, substrates, etc. The only thing that an organism seems to do with its water is to keep it constant. In recent years though, the use of X-ray, infrared, and nuclear magnetic resonance analysis in the study of the state of water in biological systems, as well as the study of its freezing pattern, the kinetic of its movement into the cell, and the observation that D_2O and T_2O do not fully substitute for water, have called the attention of biologists to the possibility that water could play an active role in biology. But today we are very far from understanding the state of water in biological systems, for one thing: physicists are not yet sure which is the structure of *pure* water itself (Frank, 1965). Reviews of water structure and properties specially suited for biologists can be found in Podolsky (1960) Klotz (1962) Kavanau (1965) Luck (1964) and Berendsen (1967) (see also Fogg, 1965). Here we will only mention some of its properties, so that we could discuss not only its role in membrane structure, but also in all those membrane phenomena considered throughout the book.

Figure 1–4 depicts a molecule of water. The two protons are at 0.965 Å of the center of curvature of the oxygen and at an angle of 104° 31′ to one another. The water molecule possesses a strong permanent dipole moment (1.76×10^{-18} esu) as well as a high polarizability (1.44×10^{-24} cm). The strong permanent dipole moment results in a very high dielectric constant (78.5 at 25 °C) which in turn shields the anion and the cation fields of the electrolytes and is one of the factors which confers to water the ability

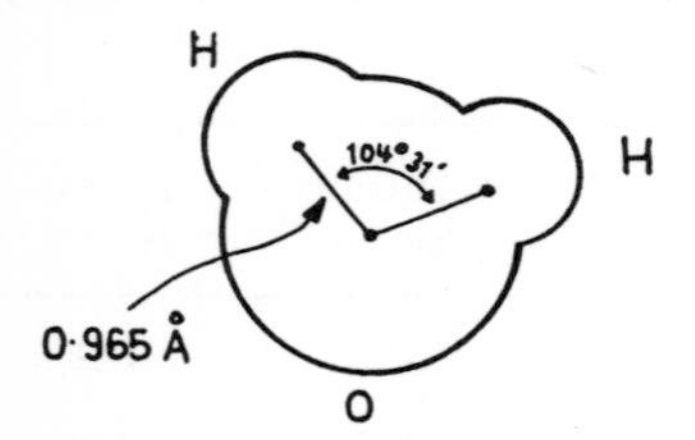

FIGURE 1–4 Water molecule.

to act as an electrolyte solvent. The charge distribution of the water molecule permits to each of the two hydrogens (low electron density) to interact with the oxygen (high electron density) of other molecules of water as depicted in figure 1–5. This interaction which is not electrostatic nor covalent, but an intermediate interaction is called *hydrogen bond* (see Orgel, 1959; Pauling, 1960; Pimentel and McClellan, 1960). It is not an exclusive property of water molecules. Whenever hydrogen is covalently bound to a strong electronegative atom (F, O, N) it can hydrogen bind another electronegative atom. Water molecules contain both, the required donor and acceptor characteristics. Even when the hydrogen bond is a weak bond (4.5 kcal/mole), there are so many water molecules participating in the membrane structure that hydrogen bonds constitute an important factor in the protein-lipid association (Vandenheuvel, 1965).

By hydrogen binding each other, water molecules can form crystal structures which, at physiological temperature, are broken down to a liquid state. However, even at these temperatures water has a considerably degree of organization (Kavanau, 1965; Bernal, 1965; Berendsen, 1967). Frank and Wen (1957) suggested the idea that water molecules hydrogen bind to each other to form clusters, or ice-like structures of ephemeral existence which, according to different models, can contain up to 28 molecules (Némethy and

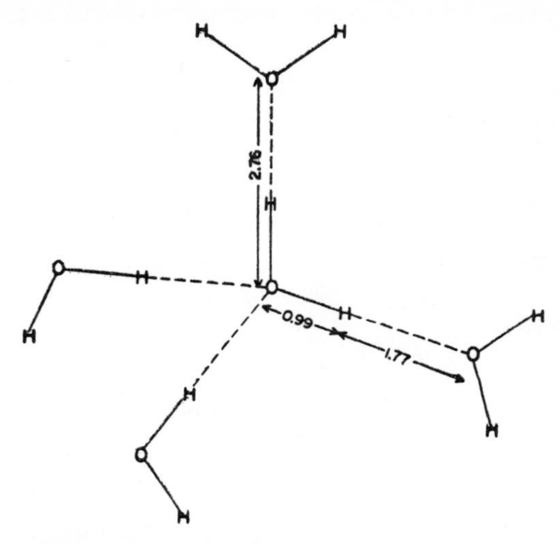

FIGURE 1–5 Hydrogen bonds (broken lines) between neighbour molecules of water.

Scheraga, 1962) or 600 molecules (Luck, 1964). On the basis that, if in a given moment a large fraction of water has ice-like structure it would have a lower density that it really has, it was argued that the ice-like structure must be very imperfect having, for instance, water molecules in the interstice of the ice-like lattice (Danford and Levy, 1962; Berendsen, 1967) or other defects. The state of water in the membrane and in the solutions in immediate contact with the membrane depends largely on the nature of the molecules forming the membrane, their distribution, and the shape of the membrane.

a) Water around non-polar groups When two separated alkyl chains are brought into water, they generally permit an organization of water around them in a clathrate-like structure with a considerable decrease of entropy (light shaded area, fig. 1–6). Therefore the system will tend to increase its entropy by expelling the two chains from the water. If they cannot be expelled, and remain in the water (say because the two chains are part of an overall water soluble molecule) then they would try to get close together so that the exposed surface of the pair will be smaller than the sum of their individual surfaces, and the amount of water that can be organized around the pair will be smaller than the sum of organized water around each separated molecule. In other words, if the two molecules attach and stay together,

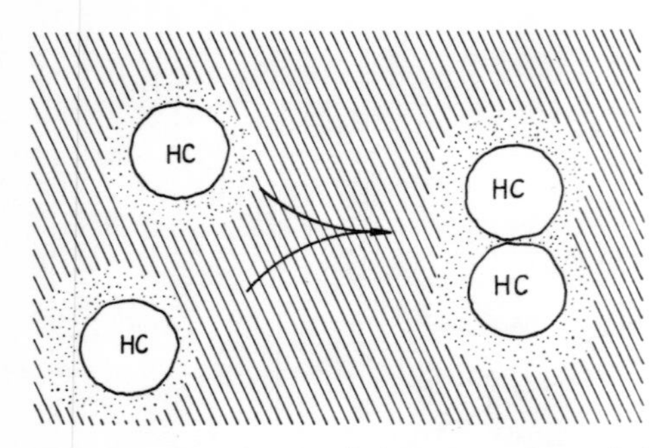

FIGURE 1–6 Schematic representation of the formation of a hydrophobic bond between two hydrophobic chains (HC). Organized water is represented by a lighter area.

they keep desorganized the water that would have been organized between them. This is the so-called *hydrophobic bond* whose thermodynamic basis were treated by Némethy and Scheraga (1962) and Némethy, Steinberg and Scheraga (1963).

b) Water in contact with non-polar surfaces The clathrate type structure described above cannot be formed if the radius of curvature of the hydrophobic surface is larger than 3.5 Å. Therefore, water in contact with a flat hydrophobic surface is not clathrate type, but has probably the same structure as in a water/air interface.

c) Water in contact with a hydrophilic surface Surfaces that can form hydrogen bonds with water can be roughly divided into two groups: those in which the distribution of hydrogen binding sites fit the lattice of the ice-like structures adopted by pure water (Fig. 1–7), and those that do not.

The first group affords, so to speak, a support on which several layers of water can be arranged. These ice-fitting surfaces are therefore likely to be covered by an extense ice-like lattice.

d) Water near mobile and fixed ions A hydrate ion consists of a core whose radius is approximately that of the crystalline compound, surrounded

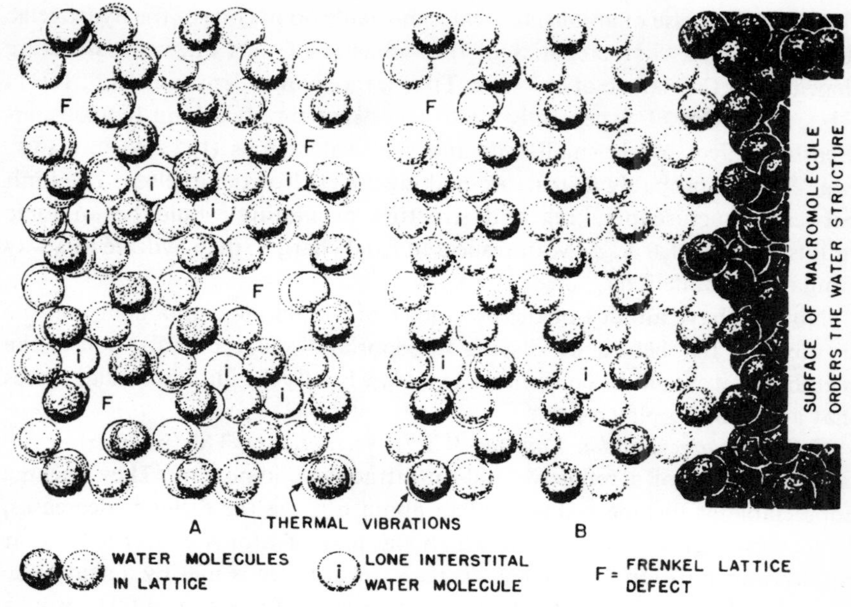

FIGURE 1–7 Stabilisation of water lattice by a macromolecular surface through hydrogen bonds (Taken with kind permission from B. Jacobson, 1955).

by a cluster of one or more layers of bound water molecules in rapid exchange (Wiegner and Jenny, 1927; Jenny, 1932; Bungenberg de Jong, 1949; Taube, 1962). The whole assembly possesses many degrees of rotational motion. Strong electrolytes like NaCl, are completely associated when in vapor state because the entropy that would be gained through dissociation is not enough as to offset the attraction between Na^+ and Cl^-. In aqueous solution, though, ions hydrate and become polyatomic ions with a much larger number of rotational states. The increase in rotational entropy of the hydrated ion together with the decrease of coulombic interaction, due to the shielding of the fields of anions and cations by the water molecules, permit the complete dissociation of Na^+ and Cl^- (see Fowler and Guggenheim, 1956). The biological relevance of this phenomenon will become evident in chapter 9 when we discuss the state of ions in the cytoplasm.

Since the interaction of water molecules among themselves is weaker than their interaction with ions, the structure of water is disturbed when ions are added. The molecules of water around a cation point their oxygen toward the ion. In the case of anions, the water molecule points one of the hydrogens. The extent and characteristics of the influence of ions on water structure depends on the nature of the ions. Thus large monovalent ions (K^+, NH_4^+, Rb^+, Cs^+, Br^-, I^-) generally have a net structure-breaking (entropy increasing) effect, while small ions and multivalent ions (Li^+, Na^+, Ca^{++}, Mg^{++}) have probably a structure-making effect (Kavanau, 1965). The limit between structure breaking and structure promoting would be an ionic radius of about 1.6 Å. As with respect to fixed charges, they hydrate in a way similar to mobile ions.

Now we have an idea of the properties of the major components of the membrane, but before we discuss the supramolecular organization of the membrane it would be convenient to review briefly the intermolecular forces that are likely to play a role.

a) Short range London-Van der Waals interactions This interaction (V) is the resultant of a repulsive and an attractive component. They become appreciable at distances shorter than about 6 Å. As the distance decreases, V goes through a minimum at which the pair of atoms concerned are at equilibrium position. A further decrease in distance is met by a sharp rise in the repulsive component. The forces between a single pair of CH_2 groups for instance, are small: only 0.002 kcal at a distance of 5 Å. However, for two lipid molecules with chains of 20 carbons long at 4.17 Å, the energy is −20 kcal which is a considerable one (Salem, 1962; Vandenheuvel, 1963). Since this force depends so critically on the close approximation between the two alkyl chains, it would be maximal between two chains that could be fit closely and parallel to each other. Unsaturations produce a kink in the chain (the bound angle between single bonded carbons is 109° 28′, while the angle between double bonded carbons is 125° 27′ [Pauling, 1960]) and, as the number of *cis* double bonds in a carbon chain increases, the kinking is more pronounced, and the force of attraction to neighbour molecules through London-Van der Waals forces is weaker. Therefore, a membrane whose lipids are highly unsaturated will not be as densely packed as one with a high degree of saturation (Vandenhauvel, 1963). This is supported by the observation that both, the permeability and the fragility or erythrocyte membranes is proportional to its content of unsaturated fatty acids (Koghl, de Gier, Mulder and Van Deenen, 1960; Walker and Kummerow, 1964; Wolfe, 1964). For further discussion see Yos, Bade and Jehle (1957).

b) Electrostatic force This force is attractive or repulsive depending on whether the charges of the two ionic groups are of different or identical sign. This force decays relatively slowly with distance. A positively charged amino group in a protein and a negatively charged phosphate in a lipid at 5 Å attract with a force of 4.1 kcal/mole.

c) *Hydrogen bond* (see water).

d) *Hydrophobic bond* (see water).

1.2 Natural and artificial preparations used in membrane research

The jump from the molecular level to the level of organization of cell membranes requires information which does not arise strightly forward from the simple combination of molecular properties. This information is being obtained through the use of some special preparations and techniques that we will now describe briefly.

1.2.1 Red cell ghosts

They were already mentioned in this chapter in connection with the membrane extraction problem. The ghosts are red cells whose hemoglobin content was removed. One of the ways of removing the hemoglobin is by swelling the erythrocytes in hypotonic buffer solutions. This procedure loosens the framework of the membrane and permits the hemoglobin to escape. Since hemoglobin constitutes the major component of the erythrocyte, the remaining material is mostly membrane. The composition of this product depends on the method of lysis and the pH used (Waugh and Schmitt, 1940; Ponder, 1952). Ghosts provide one of the most purified membrane extracts and are a source of valuable information on the thickness, chemical composition, and enzymatic activity of membranes. The ghost possesses also many of the permeability characteristics of the intact cell (Teorell, 1952; Stein, 1956; Le Fevre, 1961). The same loosening of the membrane which permits the escape of the hemoglobin permits also the penetration of substances present in the bathing solution so that ghosts provide one of the few plasma membranes in which the composition on the two sides can be controlled.

1.2.2 Monolayers

If a small droplet of diluted solution containing a known amount of lipid in a volatile solvent is deposited on a water surface, it spreads and the solvent evaporates leaving the lipid molecules scattered on the surface. Each molecule of lipid deeps the polar group into the water and leaves the hydrophobic chains in the air-phase (fig. 1–8). The behaviour of the lipid monolayer on this surface can be studied with a Langmuir-type trough. It consists of a shallow container with a bar that may be driven by a high precision screw so that the molecules of lipid are compressed against a

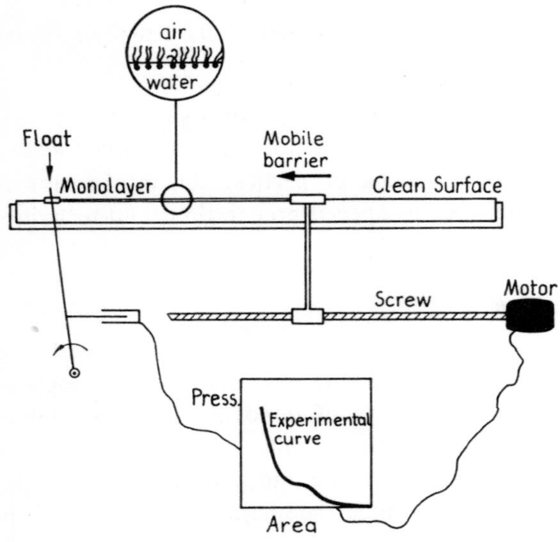

FIGURE 1–8

floating mobile barrier which is part of a pressure gauging system. The movement of the screw can be converted into area still available for the lipid layer. This area and the surface pressure developed are recorded. Figure 1–9 shows curves obtained with three different lipids. From the number of molecules deposited, and the area of the trough between the mobile bar and the floating barrier the area occupied per molecule at a given surface pressure is calculated. Other parameters, besides surface pressure/area characteristics can be measured. The surface potential can be measured by placing an electrode in the water subphase and another one above (but not touching) the monolayer. The electrical contact between this electrode and the monolayer is made by ionizing the air between the electrode and the monolayer with an alpha-ray emitter like Am^{241}. With the value of the surface electrical potential and the number of molecules we can calculate the dipole moment per molecule. As lipid molecules are brought together by the mobile bar, they interact with each other. The gradual slope of the pressure-area curve indicates the compressibility of the monolayer. The molecules of straight-chain fatty acids like stearic acid do not interact appreciably until they are quite close to each other, but then the pressure rises steeply and they cannot be further compressed without destroying the monolayer. If the fatty acids chains have unsaturations and are not straight (*e.g.*, oleic acid), they start interacting at a larger area per molecule. If they have more unsaturations or a net charged polar groups they start interacting at even larger intermolecular distances. The position and interactions of the polar groups can be inferred from surface potential studies.

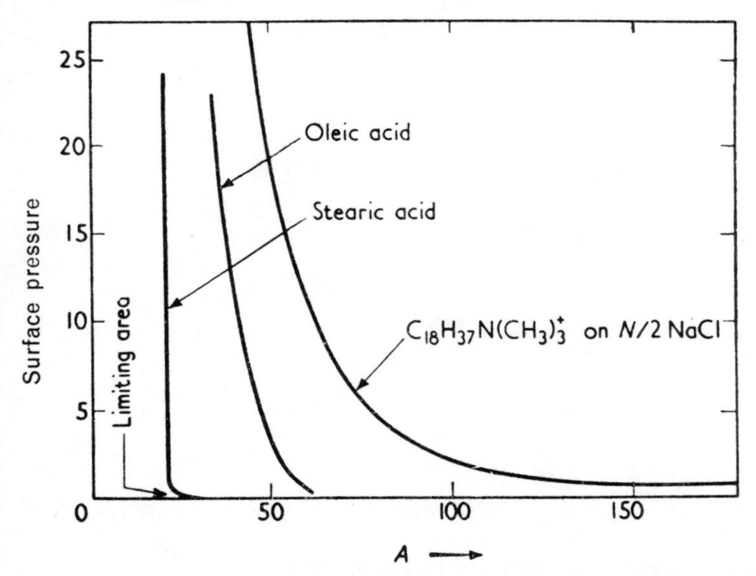

FIGURE 1–9 Surface pressure/area relationship for three different lipid species as studied with the device depicted in figure 1–8. *Ordinate:* surface pressure (π) in dyn.cm^{-1}. *Abscissa:* area (A) per molecule in Å^2. (After, J.T.Davies and E.K.Rideal, 1961.)

Lipids like those existing in biological membranes, with more than one fatty acid chains, double bonds, and bulky polar heads containing two or three fixed charges, exhibit a complicate interaction both, among themselves and with the water and ions in the subphase. Figure 1–10 shows the curves of surface pressure (π), surface potential (mV) and surface dipole moment (μ_1)

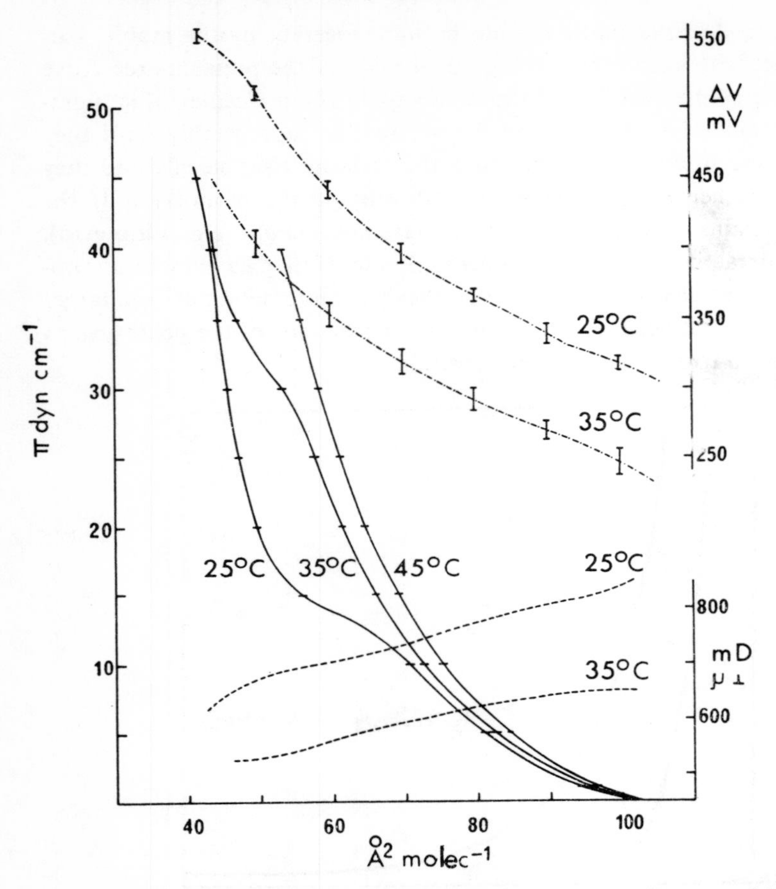

FIGURE 1–10 (— —) surface pressure (π); (-.-.-.) surface potential (ΔV); (-----) dipole moment per molecule (μ_1) of L-α-dipalmitoyl lecithin (from Vilallonga, Fernández, Rotunno and Cereijido, 1969).

per molecule *vs*. area per molecule of L-α-dipalmitoyl lecithin (DPL) at different temperatures obtained by Vilallonga, Fernández, Rotunno and Cereijido (1969). As the surface is decreased, the alkyl chains tend to arrange themselves parallel with respect to each other. The polar group is fully charged at the pH used and the positive trimethylammonium group seems to act as a "free" counterion of the comparatively fixed phosphate group of nearest neighbour molecules (Parsegian, 1967). This would give rise to bidimensional arrays of the type proposed by Pethica (1965) and depicted in figure 1–11. The decrease of the slope between 68 and 58 Å^2 per molecule in the surface pressure curve of 25 °C corresponds to the distance between DPL molecules at which the Van der Waals forces between alkyl chains and the electrostatic forces between opposite charges in neighbouring molecules reach a maximum value and the lattice has its lowest energy level. At 35°C the increased thermal fluctuation makes it necessary to compress the monolayer to an area smaller than 60 Å^2 per molecule before the change in slope is

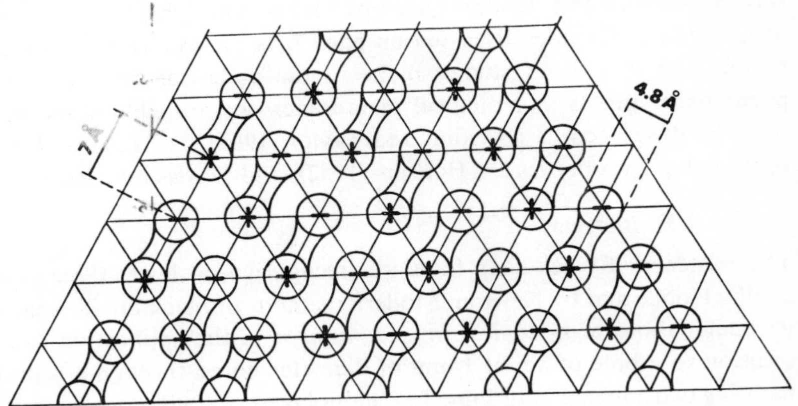

FIGURE 1–11 Bidimensional array of the polar groups of the molecules of L-α-dipalmitoyl lecithin at the air-water interface as seen from the water phase.

reached (Vilallonga, 1968). Finally at 45 °C the thermal energy is so high that there is no surface pressure (in the studied range) at which the lattice could be stabilized. The drop in surface potential with temperature would indicate that the positive choline groups involved become loose and can deep into the aqueous phase. The rise in surface potential and drop in total normal dipole moment per molecule as the area is decreased, would suggest that the contribution of the dipole formed by the carbonyl of the ester link, to the total dipole moment, decreases as they are forced to a horizontal position. Monolayers techniques not only permit the study of the behaviour of the film of a lipid species, but also the interaction of the different lipids (Dervichian, 1958), on the interaction with proteins (Schulman and Hugges, 1935; Schulman and Rideal, 1937; Dervichian, 1943; Vilallonga, Altschul and Fernández, 1967), the interaction with ions in the subphase (Rojas and Tobias, 1965; Shah and Schulman, 1965 and 1967; Standish and Pethica 1968; Cereijido, Vilallonga, Fernández and Rotunno, 1968). The study of protein films also affords valuable information on the behaviour of proteins at the air-water interphase and, the more biologically important, lipid-water interphase. Polyaminoacids and proteins at an air-water interface tend to extend themselves so as to take the nonpolar side chains out of the water and keep the peptidic backbone into the water phase. They tend to do the same in an interface covered by a lipid monolayer (Neurath and Bull, 1938; Davies, 1953; Cheesman and Davies, 1954), in particular if there is low surface pressure and the lipids are not packed too tightly. The experiment of Gorter and Grendel discussed on page 7, is an example of the contribution that monolayer studies make to the understanding of membrane phenomena. Lipids in biological membranes are probably compressed at some 30 dynes $\times$ cm^{-1} (Haydon and Taylor, 1963). For further discussion of monolayer properties see Harkins (1952), and Davies and Rideal (1961).

1.2.3 Bilayers

The existence of black soap films was recognized, at least, three centuries ago by Hooke and by Newton. Modern methods of black film formation, in particular those of biological interest, consist in the application of a lipid solution to a hole of about 1 mm of diameter bored through a septum separating two aqueous solutions. Two monolayers of lipid—one at each interface—separate the bulk lipid solution from the aqueous solution. The bulk solution of lipid gradually disappears from the meniscus, the excess of lipid accumulates in a torus around the framework, and the two monolayers contact forming a bilayer membrane (fig. 1–12). Van der Waals forces hold the hydrocarbon chains together and the polar groups coordinate with each other and with water and ions in the water phase as they do in monolayers. This preparation permits measurements of the electrical resistance, electrical capacity, thickness and permeability as a function of the lipid composition, the composition of the bathing solutions temperature, etc. The electrical properties are quite similar to those of cell membranes. For recent studies in the properties of these membranes see: Muller, Rudin, Tien and Wescott (1964); Huang, Wheeldon and Thompson (1964); Huang and Thompson (1965); Hanai, Haydon and Taylor (1965); Maddy, Huang and Thompson (1966); Tien and Diana (1967); Henn, Decker, Greenawalt and Thompson (1967); Ohki and Fukuda (1967); Miyamoto and Thompson (1967).

This preparation offers the advantage that most of the different factors involved are known. It also offers the possibility of introducing substances like antibiotics or extracts of tissues that change the bahaviour of the bilayer to such an extent that they become selectively permeable to cation or anions, show excitability resembling axon potentials, etc. In chapter 6 we will study some of these phenomena in more detail.

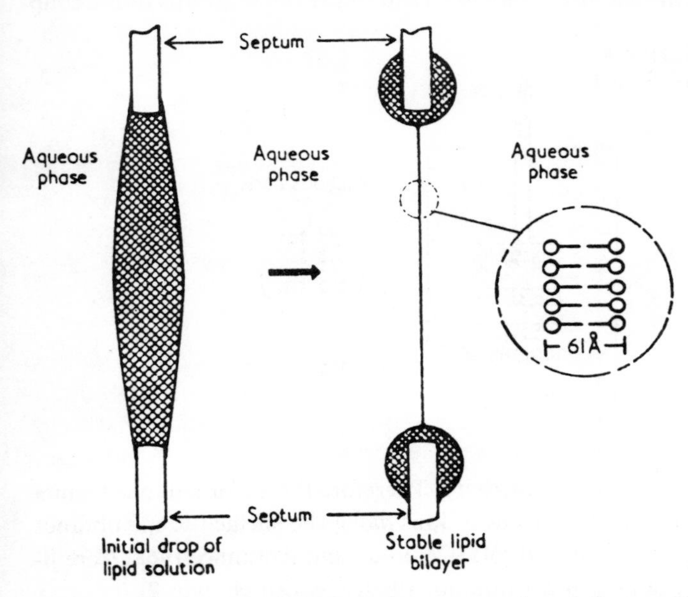

FIGURE 1–12 Diagrammatic representation of the transition of a thick lipid film to a bilayer film (Taken with kind permission from C. Huang, L. Wheeldon, and T. E. Thompson, 1964).

Clunie, Corkill, Goodman and Ogden (1967) have prepared a bilayer of C_{16}-sultaine* using the device depicted in figure 1–13. It consists of a tank containing a solution of C_{16}-sultaine 0.45 mM with or without added electrolyte (NaBr). A pair of bright platinum electrodes with a "box kite" configuration are submerged into the soap solution and then lifted so that a vertically-draining film is formed. Film thickness is calculated by measuring the optical reflection coefficient. The orientation of the molecules of lipids in the bilayer formed (fig. 1–13) is just the opposite to the one depicted in figure 1–12. Using an a.c. current Clunie *et al.* measured the conductivity *parallel* to the surface. The thickness of the liquid core was calculated using the total thick-

* 3-(dimethylhexadecylammonio)-propane-1-sulphonate (a soap).

ness of the film and the known length of the C_{16}-sultaine molecule. By studying the conductivity as a function of the thickness and NaBr concentration and comparing this value with the conductivity of the free solution, Clunie *et al.* concluded that not only the free Na^+ and Br^- in the aqueous core are mobile, but that also those adsorbed to the polar groups of the soap

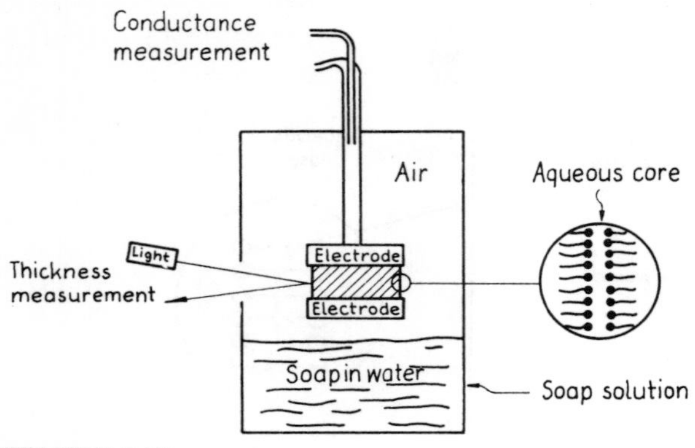

FIGURE 1–13

contribute to the measured conductivity. Therefore the adsorbed ions form a highly mobile layer. The movement of ions *along* the surface of membranes may play an important biological role (Cereijido and Rotunno, 1968; Cereijido, Vilallonga, Fernández and Rotunno, 1969; see also chapter 7).

1.2.4 Spherical lipid bilayers

This model membrane system devised by Pagano and Thompson (1967) is prepared as follows: a NaCl density gradient is formed in a rectangular glass cell (fig. 1–14). A syringe microburette attached to a narrow tube is filled with a salt solution of some intermediate density. The end of the tubing is coated with 4% egg phosphatidylcholine dissolved in a tetradecane-chloroform-methanol mixture. After placing the tube into the gradient, a droplet of aqueous NaCl is discharged and is simultaneously coated with the lipid solution. The detached droplet falls through the gradient until it becomes isodense with its surrounding at which point it comes to rest. A bilayer starts to form at the bottom and progress upwards around the droplet. The excess lipid, being lighter than the salt solution, accumulates at the top of the sphere. The sphere, of some half a centimeter in diameter, can be impaled from the top with glass micropipets and the electrical properties (membrane resistances, breakdown potentials, current/voltage characteristics) can be studied. The properties of this system are quite similar to those of planar bilayer membrane systems. The sphere, however, has a much larger area and facilitates the study of transport phenomena as a function of the composition of the inner and outer solution. Since the sphere is not attached to a mechanical support it is more resistant to technical manipulations and permits electrophoretic type experiments.

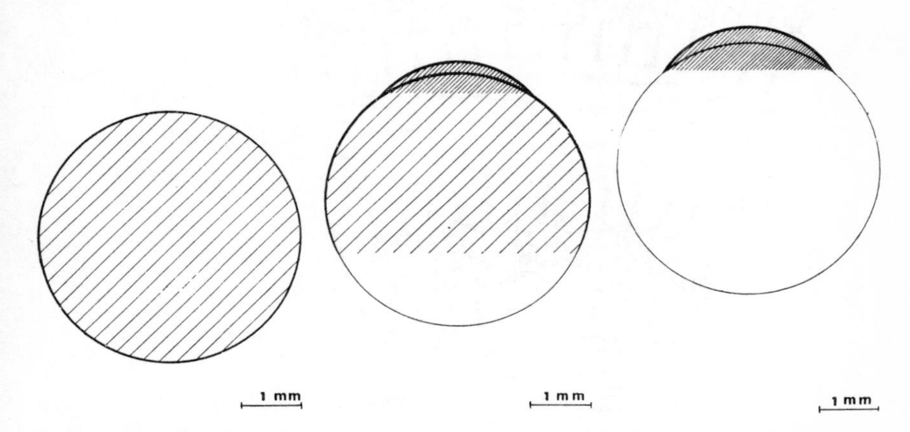

FIGURE 1–14 Three successive stages during the formation of spherical bilayer (Taken with kind permission from R. Pagano, and T. E. Thompson, 1964).

1.2.5 Myelin figures

On the basis that membranes are involved in most of the basic processes of life, that the different membranes have structural peculiarities, and that —as we shall see later—they change their structure to adapt themselves to various functions, a great effort was made to correlate the structure of membranes with their appearance in electron microscopy. Myelin figures played an important role in the understanding of the electron microscopical image of the plasma membrane. When lipids extracted from brain are put in contact with water they form smooth or twisted forms as cylindrical or disclike structures (Nageotte, 1927; Schmitt and Bear, 1937). These myelin figures are confocal structures which do not coalesce on contact with each other (fig. 1–15). Each layer is thought to be a molecular bilayer of the type depicted in figure 1–2 and 1–12 (Dervichian, 1946).

When fixed with OsO_4 and observed with an electron microscope they appear as in figure 1–16. Compare the aspect of this myelin figure with the nerve myelin shown in figure 1–19. To correlate the image with the molecular structure it is extremely important to establish where the Os is fixed.

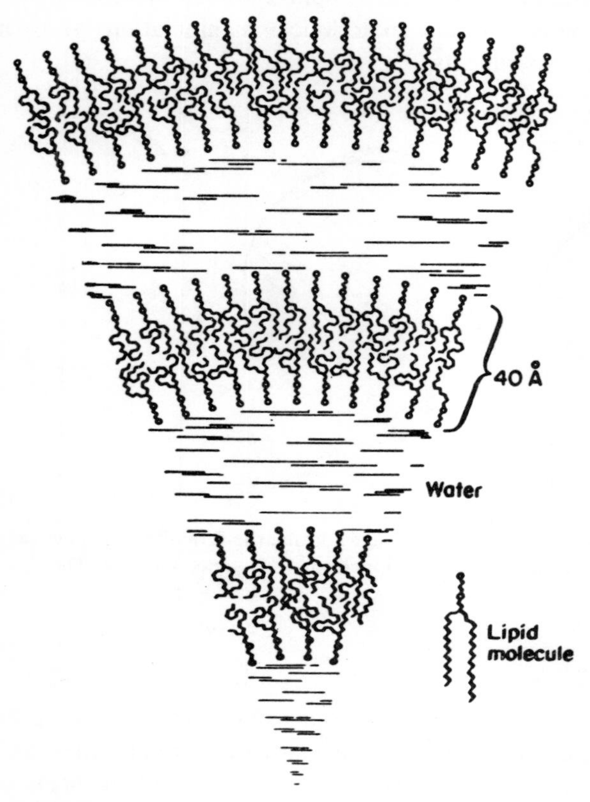

FIGURE 1–15 Arrangement of lipids and water in myelin figures (Taken with kind permission from W. Stoeckenius, 1962b).

In a series of studies correlating observations with polarized light, X-ray diffraction, and electron microscopy of myelin structures with different degrees of hydration it was established that Os is probably deposited at, or near, the polar head groups of the lipids (Stoeckenius, 1962a; see also Frey-Wyssling, 1953), so that the leaflets are considered to contain osmium in the form indicated in figure 1–17.

FIGURE 1–16 Electron micrograph showing myelin figures of brain phospholipids fixed with OsO_4 (Taken with kind permission from W. Stoeckenius, 1962b).

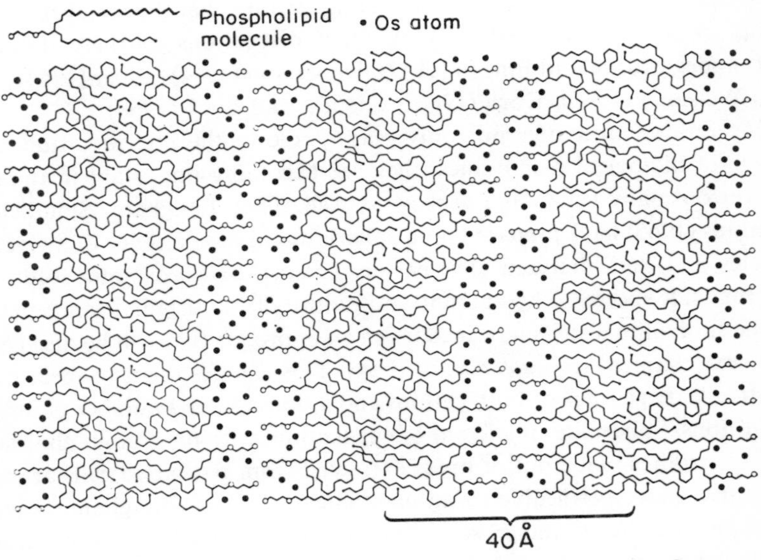

FIGURE 1–17 A possible distribution of Osmium in myelin figure (Taken with kind permission from W. Stoeckenius, 1962b).

On the basis that OsO_4 only marks phospholipids with double bonds, it was questioned whether this interpretation is correct. Stoeckenius (1963) has shown that uranyl linoleate (18 carbons, 3 double bonds), which has a dense uranium atom attached to the polar group, gives a similar electron microscope image both, before and after exposure to vapors of OsO_4. He reasoned that if osmium were fixed to the double bonds of the hydrocarbon chains, the image when exposed to the vapors would have bands due to the unranium fixed to the polar groups and bands due to osmium fixed to the double bonds. The fact that he observed no difference between the images before and after the exposure to OsO_4 is taken as an indication that Os attaches to the same site that uranium does, *i.e.* the polar groups of the lipids. Stoeckenius, 1962b, has also shown that when myelin figures are put in contact with a protein solution, the protein is absorbed on the outermost layer of lipid. Upon treatment with OsO_4 the layer covered with protein appears thicker than those with lipid only. This suggests that the image of the cell membrane, as seen in electron microscopy, is given by the contribution of the lipids and the protein as well.

As we shall discuss below, lipid layers are just one of many structures that lipids can form. Therefore, an effort was also made to device electron microscope techniques in which other configurations that the membrane might adopt could be preserved. For excellent discussions of these points see Fernández Morán and Finean (1957), Robertson (1964); Elber (1964); Korn and Weisman (1966), Korn (1966); Glauert and Lucy (1968).

Myelin-like structures not only play a role in the interpretation of the image of biological membranes, they are becoming important in the elucidation on many functional aspects as well. Thus Bangham, Standish and Watkins (1965) have prepared lipid spherulites in the presence of a given salt species and then dialyzed exhaustively against an iso-osmotic solution of another salt species. The amount and nature of the original salt remaining in the spherulite were determined using an approach that will be discussed in the next chapter (see *wash out*, page 54). This enables them to measure the relative leakage rates of both cations and anions under a number of different conditions. Spherulites made out of egg licithin, with or without cholesterol, and spherulites made out of lipids extracted from red blood cells are about 10^5 times more permeable to anions than to cations (Bangham, Standish, Watkins and Weissmann, 1967). The amount of solute contained between the lamellae varies as a function of the potential at the phospholipid–water interface: if the lipids have a net charge, as the ion strength of the solution increases the volume of the liquid compartment decreases. This effect is not observed if the lipids used have a zwitterionic polar group (*e.g.*, phosphatidyl choline). The spherulites are permeable to water and behave as almost perfect osmometers when alkali metal salts, glucose, sucrose or mannitol are used as solutes. Other solutes show graded permeabilities (Bangham, de Gier and Greville, 1967).

As we will discuss in detail in chapter 6, there are antibiotics which increase the permeability to ions of both, natural and artificial membranes. Sessa and Weissmann (1966) have found that these antibiotics (Nystatin, amphotericin B) have essentially the same effect in lipid spherulites.

1.2.6 Nerve myelin

Schmidt (1936, see also Schmitt, Bear and Palmer, 1941) demonstrated that the lipoproteins in the nerve myelin are not randomly distributed but they are highly oriented. It was later shown by Geren (1954) and Maturana (1960) that the myelin sheath consists of Schwann cell membranes wrapped around the axon of the neuron. Figure 1–18 shows four successive stages in the process of myelin formation in which the cell membrane of the Schwann cell forms a mesaxon (2 juxtaposed cell membranes) spiralling again and again around the axon. This process, together with the disappearance of the cytoplasm of the Schwann cell from in between the successive loops of adjoined membranes, originates the structure that can be observed in figure 1–19. Compare this structure with the myelin figures shown in figure 1–15 and 1–16. This particular arrangement of cell membranes is ordered enough as to permit the use of X-ray diffraction techniques. The low-angle equatorial reflections can be analyzed by Fourier analysis yielding curves which represent the electron-density distribution through the thickness of the myelin layer. Figure 1–20 shows the curves corresponding to a sciatic nerve of the rat obtained by Finean (1962) together with the interpretation of the molecular arrangement of the lipids. The highest electron-densities are thought to be given by the phosphate groups, while the hydrocarbon region gives the lowest. The minor troughs in the electron-density curves are probably due to the non-lipid components of the membrane. Notice that in a given unit membrane, one lipid leaflet gives a higher peak than the other. This is probably due to different numbers of phosphate groups in the two halves of the bilayer, which in turns reveals the molecular asymmetry of the membrane. This interpretation is in keeping with conclusions drawn from electron microscopic studies (Finean, 1960; Robertson, 1960 and 1964).

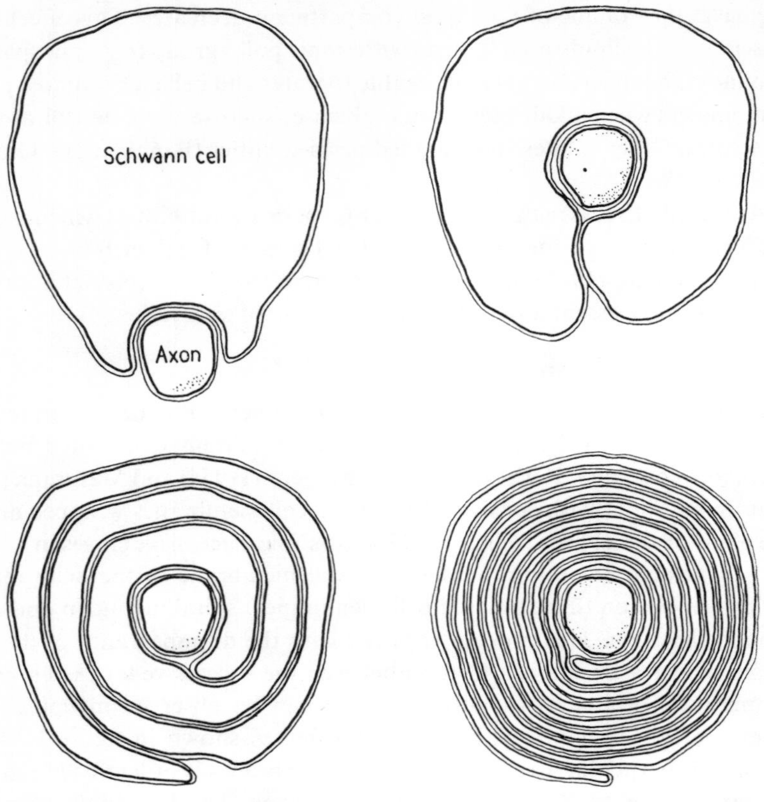

FIGURE 1–18 Four stages in the formation of the myelin sheath (Taken with kind permission from F. O. Schmitt, 1959).

Vandenheuvel (1963, 1965) has postulated a model of nerve myelin that we will discuss briefly because it illustrates a somewhat different approach to membrane studies. He classified myelin lipids into three groups: *a)* cholesterol; *b)* glycerophosphatides and *c)* sphingolipids. Based on Dervichian's observation (1958) that the area covered by a monolayer of a mixture of cholesterol-phospholipid is smaller than the sum of the individual areas occupied by a cholesterol monolayer and a phospholipid monolayer, Vandenheuvel postulates that cholesterol forms complexes with lipids of the groups *b* and *c*. The two kind of complexes are: I. *cholesterol-glycerophosphatides complexes* (fig. 1–21); they are some 30 Å long and have a cross section area of 100 Å^2. The unsaturations in chain 2 are always beyond carbon 9, *i.e.*, the

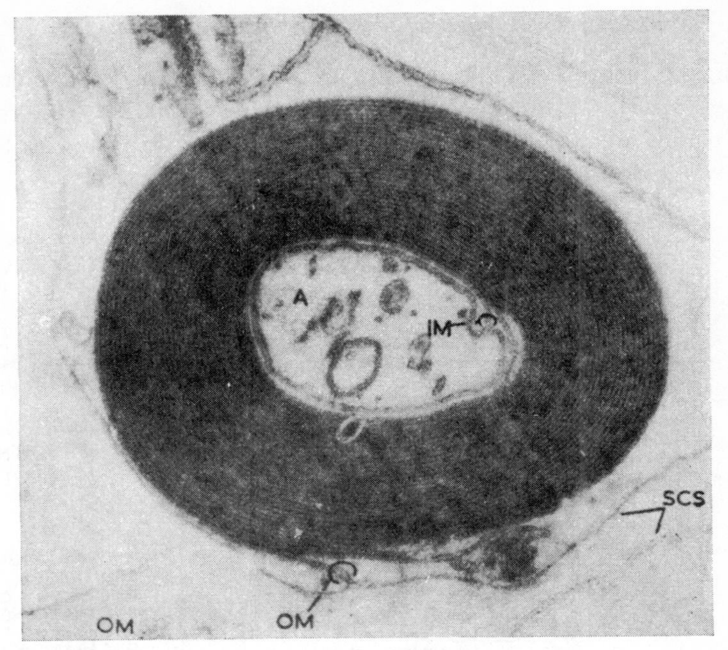

FIGURE 1–19 Electron micrograph showing a cross section through a small, myelinated nerve fibre. A: axon; IM: inner mesaxon; OM: outer mesaxon; SCS: Schwann-cell surface (Taken with kind permission from J.B. Finean, 1962).

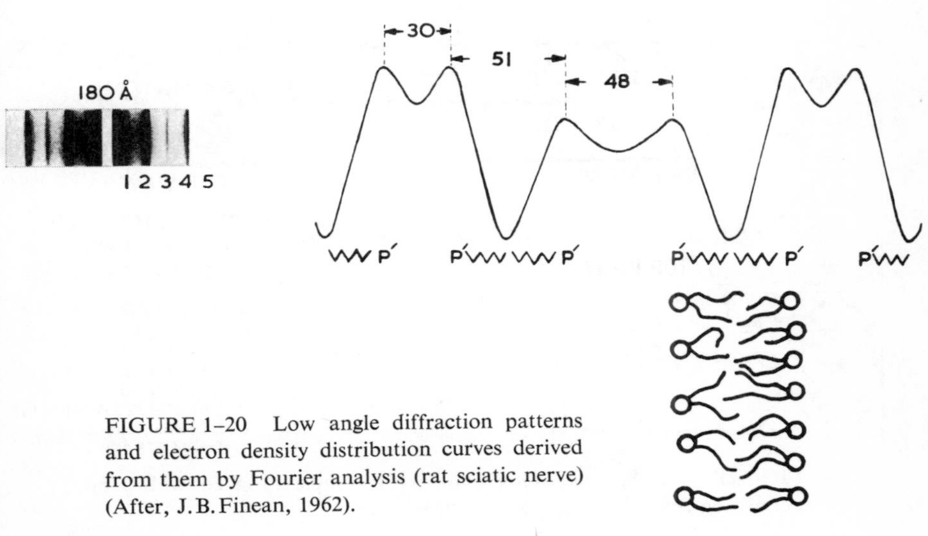

FIGURE 1–20 Low angle diffraction patterns and electron density distribution curves derived from them by Fourier analysis (rat sciatic nerve) (After, J.B. Finean, 1962).

chain only curves after C_9 thus leaving room for the complexed molecule of cholesterol. II. *Cholesterol-sphingolipid complexes* (fig. 1–22). The cross section of this complex is also 100 Å^2, but it is longer than complex I: 37 Å. The two hydrocarbon chains are straight.

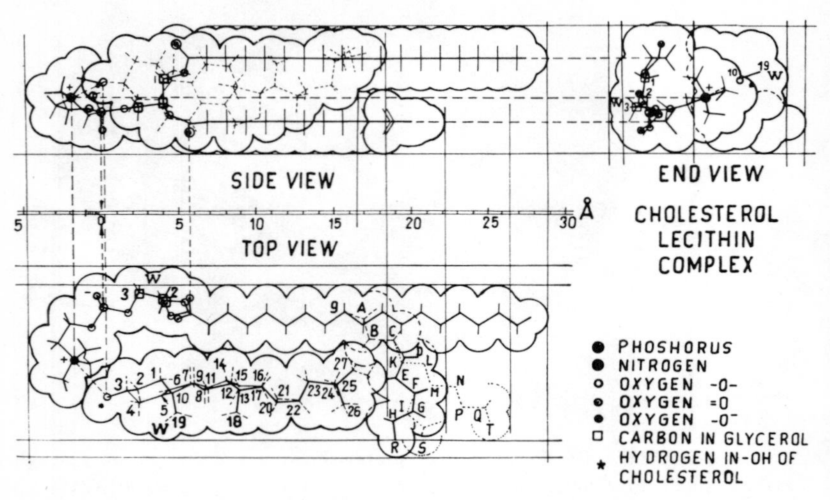

FIGURE 1–21 Cholesterol-lecithin complex (Taken with kind permission from F. Vandenheuvel, 1965).

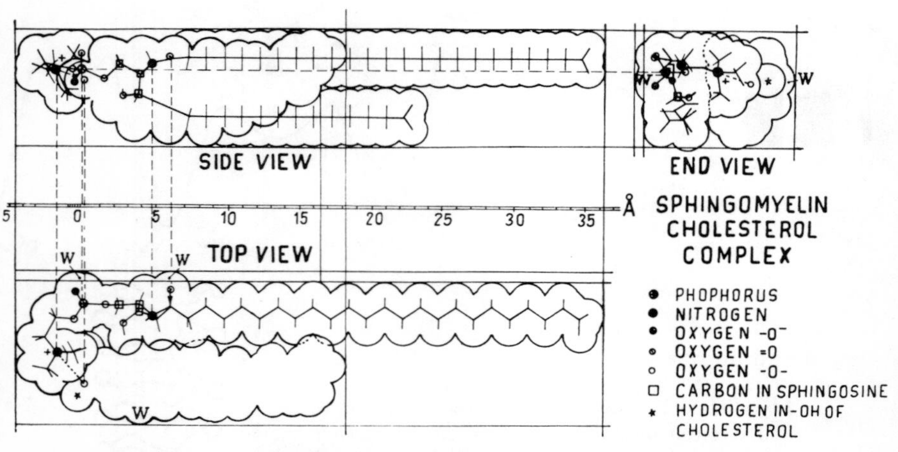

FIGURE 1–22 Sphingomyelin cholesterol complex (Taken with kind permission from F. Vandenheuvel, 1965).

When two complexes are arranged tail-to-tail as in figure 1–23 so that, according to X-ray diffraction studies of the myelin, there are 51.5 Å between the carbons of galactoses or between phosphorus atoms, one notice that: *a)* the tails of complexes type I barely touch each other, *b)* the tails of complexes of the type II interdigitate so that London-Van der Waals interactions attach the two complexes, *c)* fatty acids in a complex type II (chains 2)

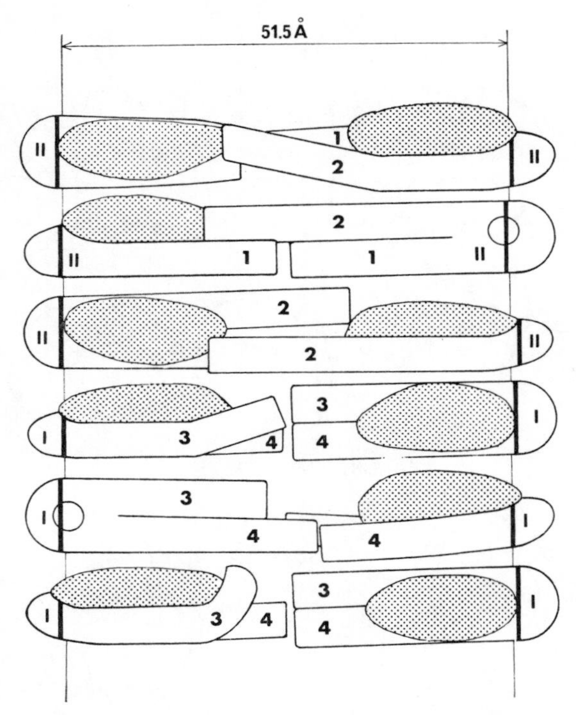

FIGURE 1–23 Possible arrangement of lipids in the myelin membrane (Taken with kind permission from F. Vandenheuvel, 1965).

are straight and establish close associations with their neighbours, thus confering great stability to the membrane, *d)* fatty acids in complexes I are curved and do not establish strong associations.

Using data of myelin composition obtained by Hulcher (1963) and Autilio, Norton and Terry (1965) Vandenheuvel tried to figure out the participation of the protein. According to these studies, in 10^4 g of dry myelin there are 2065 g of protein with an average molecular weight of 106.28 per aminoacid.

This amounts to 19.45 moles of aminoacids. The remaining component is formed by 2075 g of cholesterol (5.37 mole) which, as discussed above forms complexes of some 100 Å^2 with other lipids, and 6.94 moles of other lipids. The non-cholesterol lipids are in excess of cholesterol, so that there are some 1.65 moles of free non-complexing lipid occupying about 55 Å^2 per molecule.

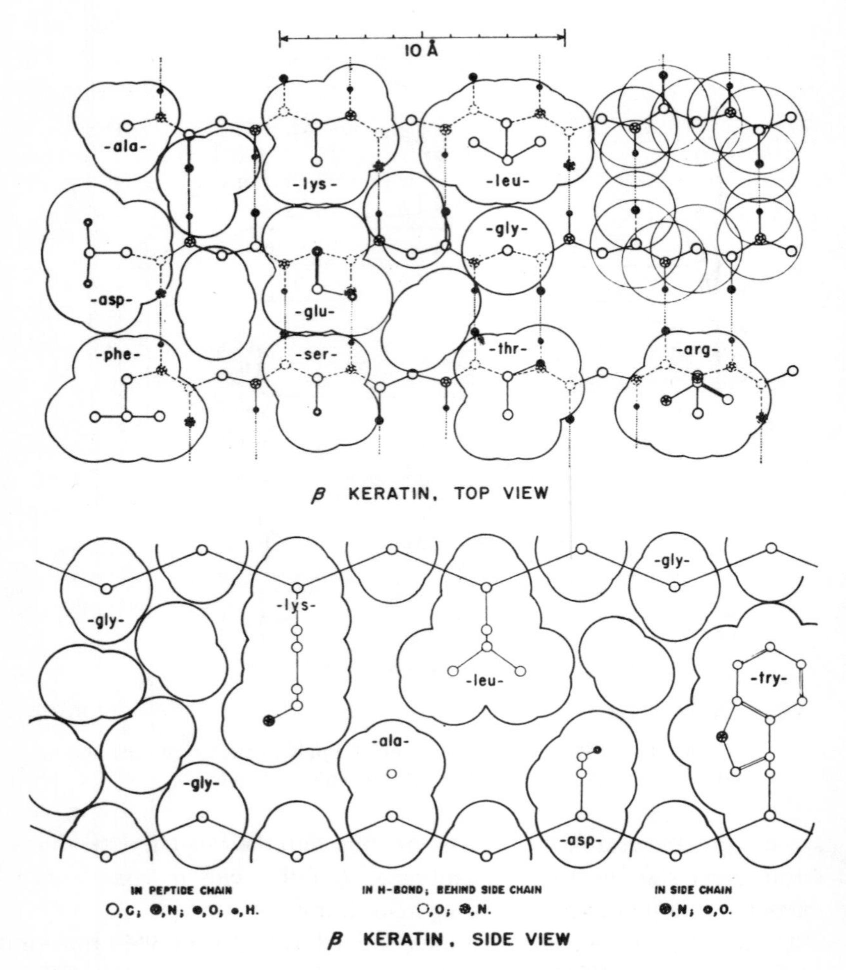

FIGURE 1–24 Van der Waals contours of amino acid side chains in two projections of β-keratin configuration (Taken with kind permission from F. Vandenheuvel, 1965).

Therefore, the area occupied by lipid is

$$6.03 \times 10^{23} [(100 \times 5.37)_{\text{complex}} + (55 \times 1.65)_{\text{free}}] = 628 \text{ Å}^2 \times 6.02 \times 10^{23}.$$

If this area is to be covered by protein, then each molecule of aminoacid has to occupy 32.3 Å^2 (*i.e.* 628/19.45). The distance between aminoacids in a peptide chain is 3.33 Å. If chains are arranged as shown in the upper part of figure 1–24, each aminoacid will cover 15.5 Å^2, *i.e.* too small an area as to account for the 32.3 Å^2 needed to cover the lipid. If chains are arranged as in the bottom part of figure 1–24 each aminoacid will cover 32.6 Å^2 (3.33 × 9.8). Vandenheuvel suggests that this is the position of the protein in the nerve myelin. He also postulates that interaction between the lipid and the protein components is based on divalent ion links (Ca^{++}, Mg^{++}) and hydrogen bonded molecules of water.

1.2.7 Liquid-crystals

The lipid bilayer depicted in figure 1–2 is just one of the many structures that the lipids can form. It follows that the structure of the membrane illustrated in figure 1–3 is one of—possibly—several conformations that the membrane can adopt. Since membrane structure and function are so intimately associated, it is of paramount importance to know what else, besides bilayers, the water-lipid systems can do. This subject, though, is too extense as to intend here a review. For excellent discussions and fundamental papers in this field see: Bear, Palmer and Schmitt (1941); Palmer and Schmitt (1941); McBain and Lee (1943); Finean (1953); Luzzati and Husson (1962); Derivichian (1964); Chapman (1965; 1966); Luzzati, Reiss-Hysson, Rivas and Gulik-Krzywicky, 1966; Luzzati and Spegt, 1967; Small (1967); Luzzati (1968). We will only refer to some aspects that could have some interest in biological systems and, among them, we will focus our attention in those that are likely to play a role in membrane phenomena.

Lipid-water systems are usually characterized by their phase diagram (fig. 1–25) which shows the properties of the system as a function of the temperature and the water content. Thus 20 per cent palmitate may be in the form of lamellas at 10 °C and in the form of micelles at 80 °C. The hydrocarbon chain can be stiff (below the T_c line) or "liquid" (above the T_c line). The conformation adopted depends on how bulky and wiggling are the alkyl chains, the size and characteristics of the polar groups, the nature of the mobile counterions that could shield the electrostatic interactions between polar groups, and the amount of water available. Luzzati (1968) (see also

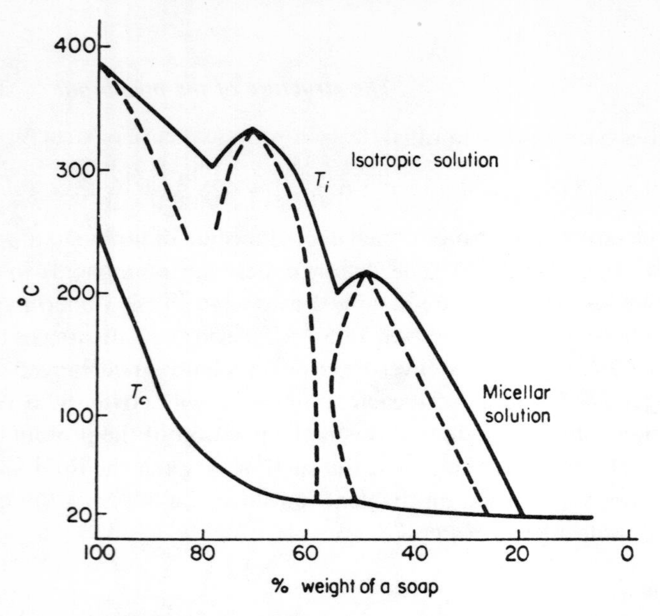

FIGURE 1–25 Phase diagrame of a lipid-water system (After J.W. McBain and W.W. Lee, 1943).

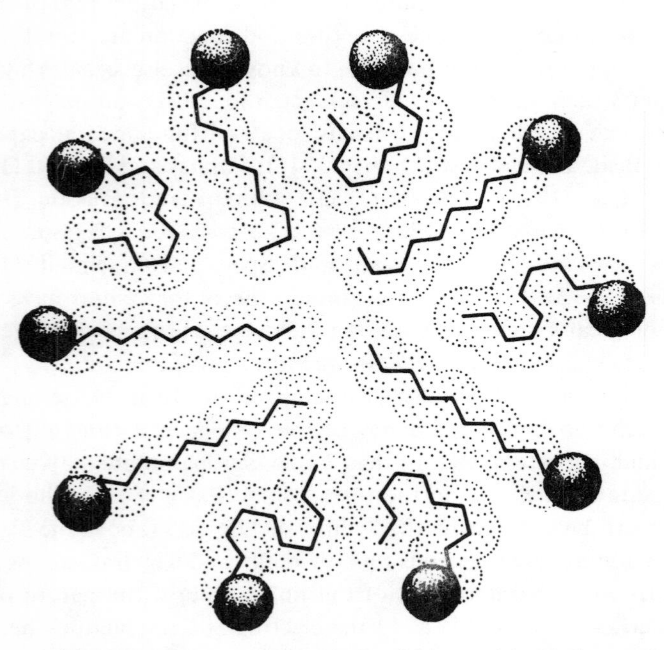

FIGURE 1–26 A lipid micelle (Taken with kind permission from J.L. Kavanau, 1965).

Luzzati and Husson, 1962) has described the following sequence of structures as the lipid concentration is rised: 1) *Micellar solution* (fig. 1–26). The lipids are in the form of spheres with the polar groups in contact with the water phase and the hydrocarbon chains in the interior. The micelle is the structure in which the polar groups occupy the relatively largest area and can be easily adopted whenever water is abundant and the polar groups constitute a large fraction of the molecule. On the contrary, when the hydrophobic chains are very voluminous, the polar group cannot hide the chains from the water and the lipid tends to abandon the micellar form and to adopt some other configuration. This was suggested to be the reason why the phospholipids, with two or more hydrocarbon chains, tend to originate bilayers.

As the concentration of lipid is rised micelles tend to elongate and to become cigar or rod shaped. 2) *Hexagonal I.* Here the lipids form long cylinders with the polar groups at the surface and the hydrocarbon chains toward the center (figure 1–27a). The area covered by the polar groups is smaller than in the micelle, is a function of the molal concentration of polar groups, and

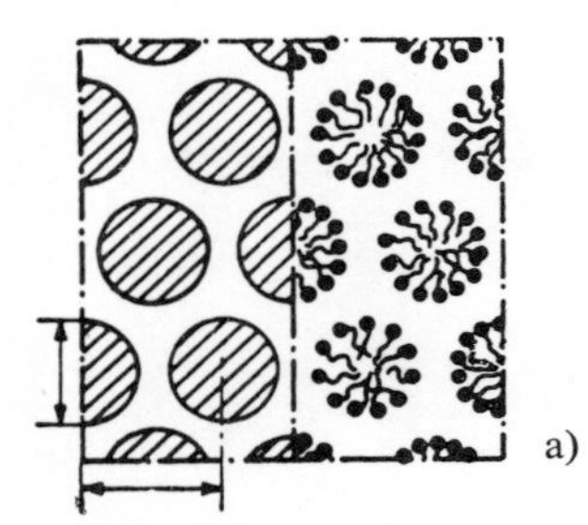

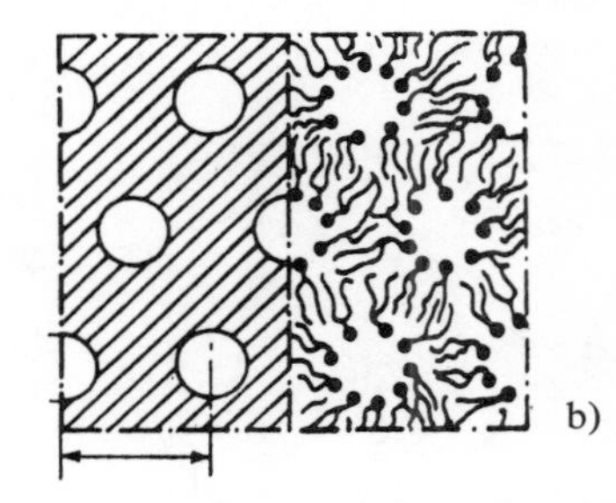

FIGURE 1–27 Cross sectional representation of *a Hexagonal I* (oil in water configuration), and *b Hexagonal II* (water in oil) configuration of a lipid water system (Taken with kind permission from V. Luzzati, F. Reiss-Hysson; E. Rivas and T. Gulik-Krzywicki, 1966).

is independent on the length of the chain. 3) *Deformed hexagonal and rectangular.* Although the structure of these phases was not unambiguously determined, the rectangular phase seems to be constituted by long (infinite) prismas as depicted in figure 1–28. 4) *Complex hexagonal* (fig. 1–29). The structure of this phase seems to consist of lipid pipes with water around and in the lumen. 5) *Cubic.* In this phase the polar groups form rods surrounded by the paraffin chain. However in this phase the rods are not indefinitely long (as in the hexagonal phase), but they fragment into segments of finite length to offer additional room for the expansion of the chains. These segments, in turn, compose two networks which interdigitate each other (Luzzati, Tardieu, Gulik-Krzywicki, Rivas and Reiss-Husson, 1968). 6) *Lamellar* (fig. 1–30). The lipid adopts the already discussed structure of bilayers separated by flat layers of water. 7) *Gel.* This is a homogeneous phase in which the paraffin

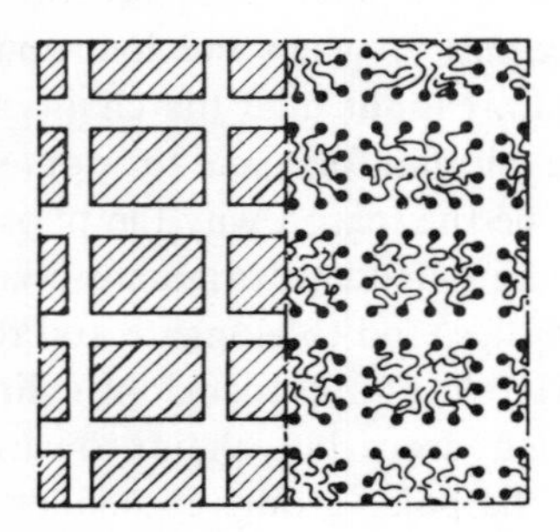

FIGURE 1–28 Schematic cross section representation of the rectangular configuration of a lipid-water system (Taken with kind permission from V. Luzzati and F. Husson, 1962).

FIGURE 1–29 Schematic representation of the complex hexagonal configuration of a lipid-water system (Taken with kind permission from V. Luzzati, and F. Husson, 1962).

chains are arranged in parallel and so tightly packed that they become straight. The same situation can be achieved at more diluted solutions by lowering the temperature below the T_c line so that the chains freeze and become stiff. The gel is thought to be a metastable state which spontaneously transforms to *coagel, i.e.* a gel which is no longer homogeneous, but in which several phases coexist in the same system. 8) *Anhydrous phases.* Here the polar groups may be segregated into lamellas in which they occupy a relativelly large area (fig. 1–31a and b). Different lipidic species do not necessarily show the same number of phases.

Liquid crystals are sensitive to the nature of the mobile ions. Soaps of divalent cations, for instance, tend to form rod-like structure (Skoulios and Luzzati, 1961). Although these structures are observed as such in chemically pure lipids, when the system has more than one lipid species it also achieves a fairly high degree of order. An interesting observation made by Gulik-Krzywicki, Rivas and Luzzati (1967) is that lipid mixtures behave somewhat differently when there is an homogeneous phase than when there are more than one phase coexisting. In the first case all lipids, regardless of chemical differences, form a unique phase. When more than one phase are

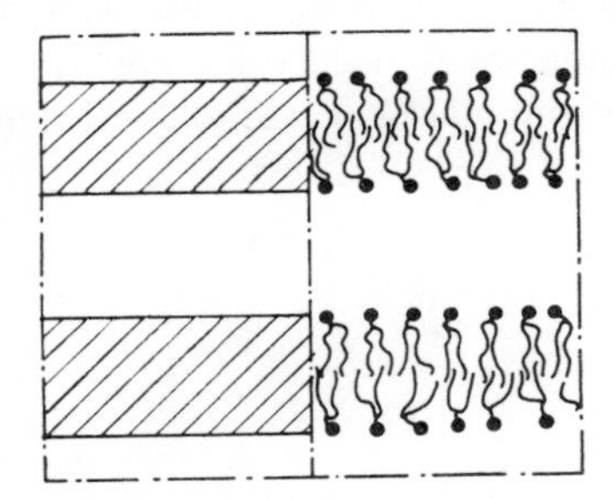

FIGURE 1–30 Schematic representation of the lamellae configuration of a lipid-water system (Taken with kind permission from V. Luzzati, F. Reiss-Husson, E. Rivas and T. Gulik-Krzywicki, 1966).

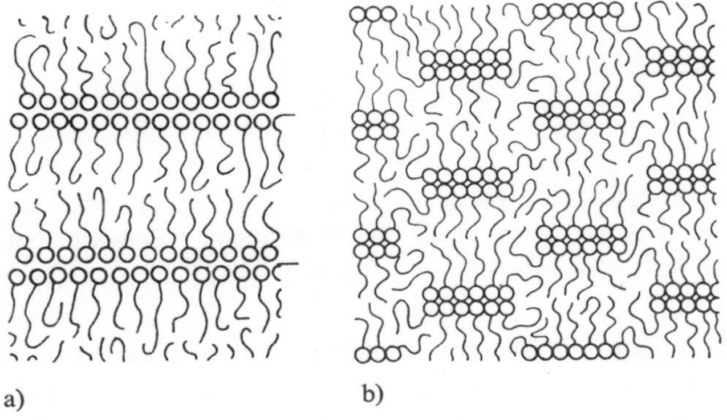

FIGURE 1–31 *a* Lamellae phase and *b* rectangular-centred phase of anhydrous soap (Taken with kind permission from A. S. Skoulius and V. Luzzati).

present, some species are segregated in a given structure and some other species agrupate to form a different structure.

When more than one lipid species are present, the transition of lipid-water systems from one phase to another is not as sharp as observed with pure components. This would suggest that the lipid component of the membrane, which consists of a wide variety of lipids may show a modulated response to a given change in the surroundings. The observation that double bonds and heterogeneity of paraffin chains make the shift gradual, lowers the temperature at which the chains become ordered, and favours hexagonal phases (Reiss-Husson, 1967) might also have a biological significance.

In connection with this, it is interesting to notice that careful electron microscope studies carried out with a negative staining technique suggest that in the cell membrane the lamellar arrangement of the lipids coexists with other structures (Doumarskin, Dougherty and Harris, 1962; Glauert, Dingle and Lucy, 1962; Glauert and Lucy, 1968).

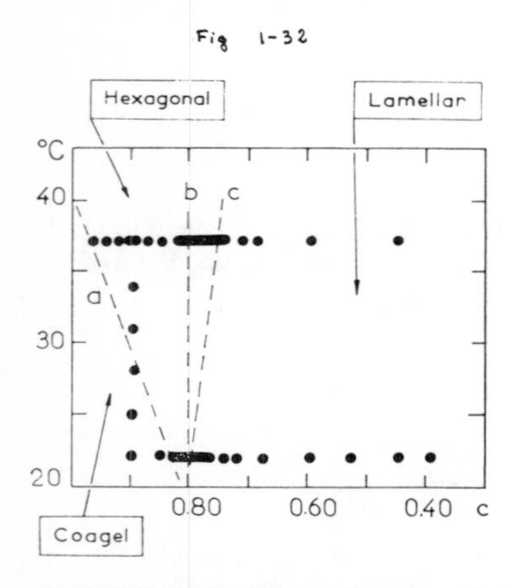

FIGURE 1–32 Phase diagram of the brain lipid water system with the positions of the experimental points (Taken with kind permission from V. Luzzati and F. Husson, 1962).

Luzzati and Husson (1962) have studied the phase diagram of a lipid extract of brain prepared by W. Stoeckenius. They observed (figure 1–32) that above line *a* (which is an equivalent of the T_c line of figure 1–25) the paraffin chains melt. Between lines *a* and *b* the structure is hexagonal. On the right hand side of line *c* the structure is lamellar. Between lines *b* and *c* hexagonal and lamellar structures coexist. The hexagonal phase observed here does not seem to be the one depicted in figure 1–27a though. Even when the spacing and intensities of the X-ray diffractions are those of a hexagonal phase, the amount of water present is not enough as to form lipid cylinders in water. The system is likely to have the structure water-in-oil *(hexagonal II)* illustrated in figure 1–27b).

Paraffin chains seem to be in liquid state in myelinated nerves (Schmitt, Bear and Palmer, 1941) but some recent nuclear magnetic resonance data would indicate that lipid chains in bacteria and erythrocytes may not be entirely in a liquid state.

Cerbón (1964, 1965 and 1967) carried out an analysis of the nuclear magnetic resonance of water and lipids in cells and artificial systems. He has shown that in these systems both water and lipid mobilities are restricted. The exposure to $CaCl_2$ results in a mobilization of the organized water and an increase in the height of the lipid signal. Cerbón suggests that lipids and water are intimately associated and immobilizate each other. $CaCl_2$, by displacing water molecules from the polar groups, would permit a higher degree of mobility.

The bearing that these lipid-water structures, and their transitions could have in membrane phenomena will become evident in chapter 4 when we discuss the possibility of pore formation, also in chapter 6 when we review the role of membranes as liquid ion exchangers, and in chapter 9 when we considerate the relationship between conformational changes and functional states.

A corollary of the observations discussed above, is that since the stability of the different lipid-water systems is so delicate, it is not likely to withstand the usual treatments for electron microscope observations. Therefore it is not surprising that, except in the very special cases of negative staining mentioned before, the only membrane structure observed in cells is the lamellar one (see Finean and Rumsby, 1963).

The self-alignment of lipid molecules in liquid-crystals and in an interface represents a minimal energy state. A change in the direction of the molecules is associated with an energy change (Frank, 1958). As pointed out by Fergason and Brown (1967), a local change in the chemical nature of some lipid molecules may result in a release of energy which, in turn, will be converted into mechanical energy in the form of a torque. The mechanical energy will cause a change of shape in the liquid-crystal. The propagation of this change will obey wave equation. In this way, the information of a local chemical change is transmitted in the same manner than a sonic wave propagates in a solid. Figure 1–33 illustrates some wave forms that might occur in cell membranes. Therefore membranes afford a way of transmitting chemical events with far more possibilities than the diffusion of a chemical "messenger" could offer.

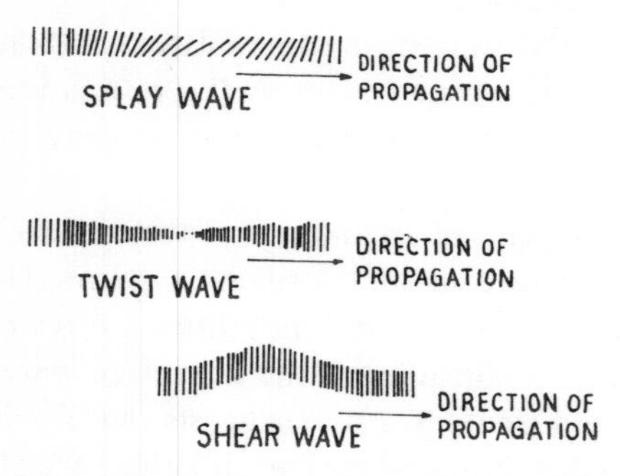

FIGURE 1–33 Torque waves in membrane-like liquid crystals (After J.L. Fergason and G.H.Brown, 1967).

At this point we have an idea of the chemical composition of the membrane, and the behaviour of its components under a variety of conditions, ranging from the situation of a lipid on the surface of pure water, to the complexity of the myelin sheath. We will now use this information to devise a general picture of the structure of the membrane.

1.3 Models of the cell membrane

The chemical analysis indicates that the different membranes have the same basic components but that their proportion varies widely. Electron microscope studies suggest an even wider difference, so do enzymatic and permeability data to be studied in the following chapters. The preparations reviewed above indicate not only that different membranes have different properties but also that the same membrane may alter both, its structure and functions according to minor changes in its composition, or the composition of its environment. The different membranous structure of a given cell also appear to have a different structure (Sjöstrand, 1968). It is therefore rather difficult—and perhaps no justifiable—to have something more than a general model of the membrane. In this respect Davson-Danielli's model (with relatively minor modifications) has been the helpful framework of three decades of membrane studies.

The lipid components adopt the bilayer or micellar configuration and could switch from one to another. The possibility exists that they form pores like those found by Luzzati's group in liquid crystals (Fig. 1–27b). They could also participate in elaborated electrical events in which their polar groups act as gates, organize water, and sort ions.

On the basis of its behaviour at interfaces proteins were thought to adopt the β-keratin configuration and form a thin film of about 10 Å on each side of the membrane. However, the amount of protein varies widely from membrane to membrane, and the possibility exists that the conformation of a protein in a membrane like myelin (where it constitutes less than one fifth of the dry residue), were completely different from its conformation in protein rich membranes. Moreover, the evidence that all proteins extend as a flat film at interfaces is by no means conclusive. Thus Malcom (1962) studied the behaviour of polymers at interfases and, on the basis of the low rate of deuterium exchange, and the infrared spectrum, pointed out that a large fraction of the polymer must be in α-helix. It was also suggested (Kavanau, 1965), that the protein may be fully extended in those areas of the membranes where the non-polar side chain can interact with the lipid component, while they adopt a random coil position where this interaction is not possible. If required by the circumstances this random coiled protein can be extended and constitute an elastic component of the membrane. Lennard and Singer

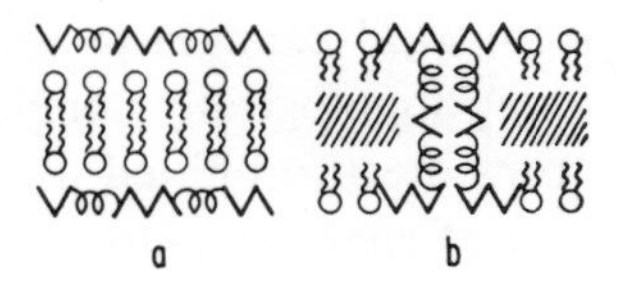

FIGURE 1–34 Schematic representation of the Davson-Danielli's model (a) and the modification proposed by J. Lennard and S. J. Singer, 1966. (b) See text (Taken with kind permission from J. Lennard and S. J. Singer, 1966).

(1966) carried out studies of the optical rotatory dispersion and circular dichroism of aqueous suspensions of membranes of human erythrocytes and *B. sublitis* which indicated that 60 to 75 per cent of the protein is likely to be in the random coil form. They suggest a modification of the Dawson-Danielli's model to account for their findings (fig. 1–34). In this model an important fraction of the protein is thought to go all the way across the membrane. This view is supported by the evidence that there is an extensive hydrophobic interaction between membrane proteins and lipids (Green and Fleischer, 1963; Brown, 1965; Wallach and Zahler, 1966). It is worth noticing that, although the model of figure 1–3 also suggests hydrophobic interactions between the protein and the lipids, they are not likely to occur in cell membranes at least for two reasons: a) lipids are thought to be packed at a pressure too high as to allow the interdigitation of the non-polar side chains of the protein. b) If the protein backbone rests on the surface of the phospholipid layer, its side chains are shorter than the thickness of the layer formed by the polar heads of the phospholipids (4 *vs.* 8 Å) thus making impossible the interaction of the side chains with the alkyl chains of the lipids. The continuity of the outer and inner layer of protein of the type illustrated in figure 1–34 was also suggested by Wallach and Zahler (1966) who envisage the protein of the membrane as consisting of two hydrophilic sections laying at the membrane surfaces and connected by hydrophobic rods which penetrate the membrane normal to its surface.

Water composes some 30–50% of the membrane and is likely to play an important role in its structure and function (Schmitt, Bear and Palmer, 1941; Finean, 1957; Fernández Morán, 1959 and 1962; Hechter, 1965a and b). Biological membranes are probably encased in a thin ice-like crust of bound water molecules which is ordered but not immobile. Since proton mobility in ice is one to two orders of magnitude greater than in water, this ice-like crust may facilitate proton transfers. As Kavanau (1965) pointed out, changes in proton potentials resulting from local changes in proton activity should be transmitted preferentially along topologically connected pathways within the crust, somewhat as electrons are transmitted along a wire. This mechanism could permit the existence of enzymatic circuits and molecular interactions with a degree of organization far above the possibilities that a conglomerate of enzymes in solution would have.

When an electric field is applied to living cells suspended in Ringer's solutions at physiological pH they generally migrate toward the anode. The position and nature of the ionogenic groups responsible for this effect have been studied by measuring the mobility of cells in an electric field as a function of pH and ionic strength both before and after the treatment with enzymes which hydrolyze the membrane molecules suspected to possess the ionic group involved (Bangham, Pethica and Seaman, 1958; Wallach and Eylar, 1961; Eylar, Madoff, Brody and Oncley, 1962; Cook, Heard and Seaman,

1962; Maddy, 1966). Not only the outermost, but also those groups located under the surface of the membrane may play a role (Hear and Seaman, 1960; Haydon, 1961). Charges in the cytoplasm, although present in a high concentration, do not seem to contribute to cell mobility in an electric field (Elul, 1967). The shape of the cell has an important influence on the field. This seems to play an important role not only as a factor affecting mobility measurements (Furchgott and Ponder, 1941), but also influencing the intercellular contacts and communication necessary to build up tissues. In this connection, it is interesting to note that cancer cells have different mobilities (different surface charges) than normal cells (Ambrose, James and Lowick, 1956; Purdom, Ambrose and Klein, 1958; Ambrose, 1964). Friedenber (1967) has published an interesting monograph on the electrostatics of cells surfaces in which many of these effects are extensively discussed.

The influence of fixed charges on ion permeability is discussed in chapter 6.

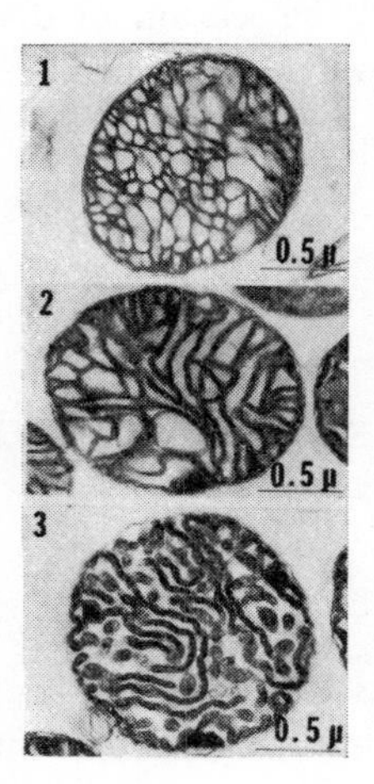

FIGURE 1–35 The conformational basis of energy conservation in membrane systems (beef heart mitochondria). 1) non-energized, 2) energized, 3) energized twisted configurations (Taken with kind permission from R.A. Harris, J.T. Penniston, J. Asai and D.E. Green, 1968).

Two places where membrane structure and function are clearly connected are the mitochondrial and chloroplast membrane (Lehninger, 1964; Deamer and Packer, 1964; Slautterback, 1965; Deamer, Utsumi and Packer, 1967). Conformational changes in mitochondria are associated to oxidative phosphorilation (Hackenbrock, 1966). The possibility exists that some of the different structures adopted by mitochondrial membranes could storage a considerable amount of energy transferable to ADP and P_i to form ATP (Boyer, Bieber, Mitchell and Szabolscski, 1966). Penniston, Harris, Asai and Green (1968) (see also Harris, Penniston, Asai and Green, 1968) have shown that the multiple configurational states of the inner membrane of the mitochrondrion are determined both by the conformation of the repeating units and also by the mode of the cristae. They classified the conformation of repeating units in non-energized, energized and energized-twisted, and the mode of the cristae in orthodox, aggregated and comminuted. They have experimentally found eight out of the nine theoretically possible combinations of these functional and conformational states. Figure 1–35 illustrates some examples of these conditions. According to their thesis an energized state can be achieved by the electron transfer (process I) or by the hydrolysis of ATP (process II). The energized state can be discharged through some of the work performed by the mitochondrion: translocation of divalent metals plus P_i; swelling and translocation of monovalent ions; or energized translocation according to the following scheme:

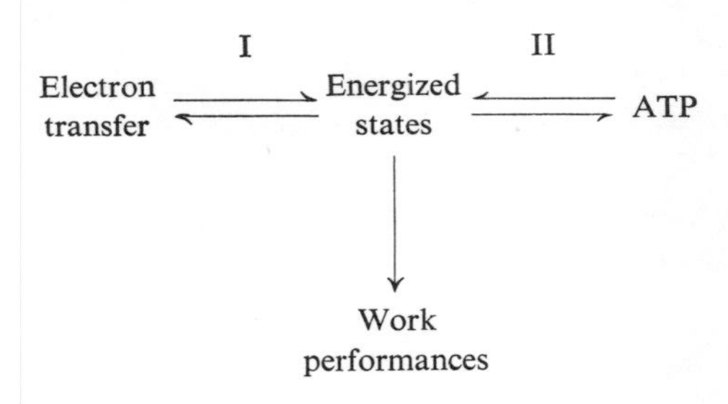

Although the discussion of mitochondrial events is beyond the scope of this book, these considerations are made because it is of particular interest to membrane biologists that the mitochondrion—an organelle specialized in the handling of energy—associated the conservation of energy with the conformational state of its membranes.

In summary, the cell membrane is a highly ordered and dynamic structure composed mainly by lipids, proteins and water, that could vary its composition and configuration, and serves as a framework for the enzymatic circuits that carry out the most fundamental processes of life. In later chapters we shall see how the membrane acts as a highly selective barrier for the passage of substances into and out of the cells.

2

Compartmental analysis

The measurement of fluxes and distribution of substances in biological systems

---

ONE OF the roles of membranes is to act as barriers between compartments. The study of how a substance moves and distributes in a system of compartments gives us information on the properties of the membranes which separate them. The amount of a substance which crosses the membrane per unit time is called its *flux* (J). Depending on whether we can obtain the necessary parameters, the flux can be expressed in micromoles per second and per square centimeter of membrane. But it is not unusual to find a flux expressed, for instance, as micromoles per hour and per kilogram of dry cells.

In a subindex, say the one in J_{12}, the number on the left (1) stands for the compartment from wich the flux comes, and the one on the right (2) stands for the compartment where the flux goes. Thus, J_{12}^{Na} means the flux of sodium from compartment 1 to compartment 2. Figure 2–1 illustrates the time course of the penetration of a substance from a large volume of solution (compartment 1) into a cell (compartment 2). The experimental conditions are arranged in such a way that: a) at the beginning ($t = 0$) the concentration of substance in the cell is zero, and b) the amount of substance in the solution is large enough so that its concentration remains constant in spite of the amount of substance which penetrates into the cell. In this case compartment 1 is said to constitute a *constant reservoir*. The flux may occur in two directions: medium → cells (J_{12}), and cell → medium (J_{21}). In the imaginary case we are discussing, the fluxes are proportional to the concentration of substance in the compartment from which they come. Therefore, at the beginning J_{21} will be zero, but as the concentration in the cell goes up,

4 Cereijido/Rotunno (0241)

J_{21} will increase. When the concentration in the cell becomes equal to that in the reservoir J_{21} will be equal to J_{12}, and J_{net} will become zero. It is very important to keep in mind the fact that, in spite of the zero net flux, fluxes in both directions do not stop when J_{21} equals J_{12}. As it will be explained in Chapters 3, 4 and 5, if we repeat the experiment with different concentrations of substance, at different temperatures, in the presence of chemically related substances, with metabolic inhibitors, etc., we could learn a great deal about the mechanism used by the substance to penetrate the cell.

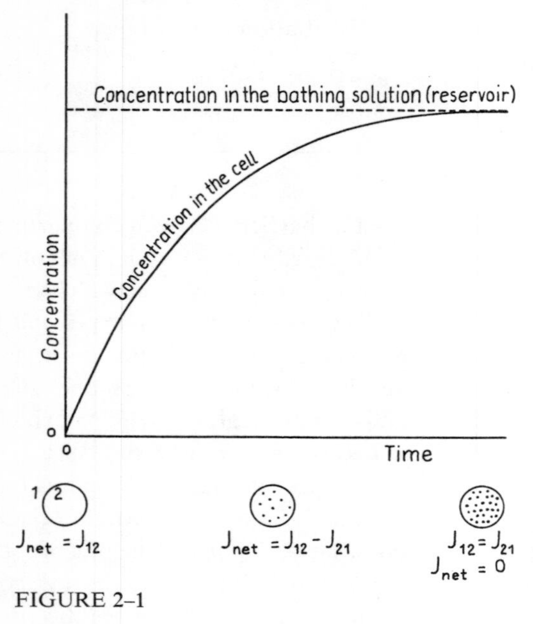

FIGURE 2–1

The conditions and experimental set ups to evaluate fluxes vary according to which parameters we can measure and where, when and how the measurements could be made. Sometimes the penetration is followed by measuring the chemical amount of substance in samples of cells taken as a function of time (see for instance Schultz, Epstein and Solomon, 1963; Miller, 1964). It could also be followed by using a substance labelled with a radioactive element and measuring the counts per minute in the cells as a function of time (see for instance Armstrong and Rothstein, 1968). If the penetration of the substance produces swelling of the cells we could use the rate of swelling to measure penetration (see for instance Sidel and Solomon, 1957).

Sometimes we have not access to the compartment where the flux goes but to the one where it originates. In this case, if we do not make of this compartment a reservoir, we could follow the penetration of the substance in the cell through its disappearance from the medium (see for instance Paganelli and Solomon, 1957). The importance of when the flux should be measured becomes evident from the experiment illustrated in figure 2–1. If we can take enough samples before J_{21} achieves a significant value, the amount of substance taken by the cell in the initial part of the experiment measures J_{12}. The interpretation of the results will also depend on whether the system is or not in steady state, whether the sampling modifies the conditions, etc.

Among the different approaches, conditions and techniques of compartmental analysis that biophysicists employ to study membranes, we will choose the use of radioisotopes in systems which are in steady state (see below) to illustrate its use. The convenience of this choice arises from the fact that many systems we deal with, do not have net fluxes, and the content of substances in the different compartments is constant. Going back to figure 2–1, we realize that, once the steady state distribution of the substance is reached (at $t = \infty$), we could add to the solution the same substance but this time in a tiny amount labelled with, say C^{14}, and measure J_{12} through the increase of counts per minute of C^{14} in the cell. To explain how it is done we need to make first some definitions and to explain some requirements:

a) Among the different meanings that the word *compartment* could be given, the one that suits best our purposes is the following: a compartment is a given *amount* of substance that from a kinetical point of view, behaves homogeneously. They can be classified into *physical* and *chemical* compartments. To illustrate this point imagine that we are studying the distribution of an aminoacid in a population of cells. Imagine also that the cells use the aminoacid in the synthesis of proteins and that this process is slower than the penetration of aminoacids into the cells. In this case there are 2 physical compartments in the system (the medium and the cell compartments) and two chemical compartments into the cells (aminoacid free and as a constituent of the proteins). Although we are interested in physical compartments because they are usually limited by membranes, we have to keep in mind the possibility of chemical compartments because of the profound effect that they have in the driving forces and equilibrium distribution.

b) The system should not discriminate between the labeled and the non-labeled molecule being traced.

c) The introduction of tracer into a system should not modify its behaviour.

d) The data used in Compartmental Analysis refering to counts per minute, specific activity, etc. are already corrected for the natural decay of the radioisotope. The disappearance of tracer from a compartment is the one due to its outflux, not to its radioactive emission.

e) The *specific activity* (p_i^*) of a tracer in compartment i is the amount (P_i) of tracer in the compartment divided by S_i, the total amount of the traced substance in the compartment.

$$p_i^* = \frac{P_i}{S_i} \qquad 2\text{–}1$$

p_{20}^* means specific activity in compartment 2 at $t = 0$
$p_{2\infty}^*$ means specific activity in compartment 2 at steady state distribution of tracer, *i.e.* after p_2^* reaches a steady value.
v_i is the volume of compartment i.
k_{ij} is a rate constant which represents the fraction of S_i which flows from compartment i to j per unit time and unit area.

$$k_{ij} = \frac{J_{ij}}{S_i} \qquad 2\text{–}2$$

p_i is the concentration of tracer in compartment i (*i.e.* P_i/v_i).

f) We will restrict our study to compartments in which the tracer distributes homogeneously throughout the compartment. In this way we can make the assumption that the only restriction to tracer movement is ascribed to the membranes, and that the time required for a substance to distribute homogeneously in the compartment is negligible as compared with the time it takes to cross the barriers. This assumption, made here for the sake of simplicity in explaining compartments, will be removed in later chapters.

g) We say that a compartment is in *steady state* when the amount of substance and the volume of the compartment have a constant value. This is a most desirable situation, because it generally occurs when the biological preparation is in healthy condition, and also because it greatly simplifies the mathematical handling of the data. As discussed in detail in chapter 3, in order to maintain a steady state the system should spend energy and, therefore, this state is different from thermodynamic equilibrium.

h) If the system is in steady state and fulfills also requirement *b*, then the driving force for tracer movement is the gradient of specific activity. Tracer movement will then follow first order kinetics regardless of the mechanism used by the membrane to translocate the tracer. As explained above, the information on this mechanism could be obtained by studying the movement of tracer with the system in different steady states (*i.e.* with different concentration of substance, at other temperatures, adding metabolic inhibitors, etc.).

i) A system of compartments can be closed or open depending on whether P_i the total amount of tracer, remains constant. If we have a beaker with a suspension of cells, it could constitute a closed system for K^{42} added to the bathing medium because the sum of K^{42} in the bathing medium (P_1^K) and in the cells (P_2^K) will be constant throughout the experiment regardless of whether the cells achieve or not steady state. The same preparation could constitute an open system for C^{14} added to the bathing medium as $^{14}CO_3HNa$ if part of it were lost as $^{14}CO_2$. In the case of K^{42}, the amount of tracer added to compartment 1 at the beginning (P_{10}) is

$$P_{10} = P_1 + P_2. \qquad 2\text{–}3$$

Since the amount of K^{42} lost by a compartment is gained by the other, the following behaviour will be observed.

$$\frac{dP_1}{dt} = -\frac{dP_2}{dt} \qquad 2\text{–}4$$

j) The system can be irreversible if one of its steps includes a boundary through which fluxes are

$$J_{ij} > J_{ji} = 0.$$

In a closed system of two compartments, for instance, if the J_{21} is zero but J_{12} is not, and the compartments are not reservoirs, the tracer and the traced substance will accumulate in compartment 2.

k) A reservoir is not necessarily constant. Imagine that a substance whose penetration into leucocytes we want to study is injected at $t = 0$ into the blood stream. The amount of the substance taken by the leucocytes will be probably negligible as compared with that contained in the plasma. The plasma will therefore behave as a reservoir. However, the amount in the plasma could decrease continuously for other reasons (penetration into other cells, renal excretion, catabolism, etc.). Of course the calculations are easier when reservoirs are constant.

2.1 *Basic experimental set ups for tracer measurements*

2.1.1 *Uptake*

Uptake is the case of a cell suspension in a beaker when the tracer is added to the bathing solution. The process can be followed by taking periodically a sample of cells. It may have two main varieties: a) Given when the amount of substance carried by the flux compares to the amount of substance in the two compartments. Figure 2–2a illustrates the time course of this experiment. For a real case in which this technique was used to measure the uptake of glucose by human red cells see Le Fevre and McGinniss (1960). b) Given in the same system when the cell suspension is very diluted. In this case the bathing solution acts as a constant reservoir. Figure 2–2b illustrates the time course. This technique is illustrated in a work by Aull and Hempling (1969) in which they study the sodium movement in tumor cells.

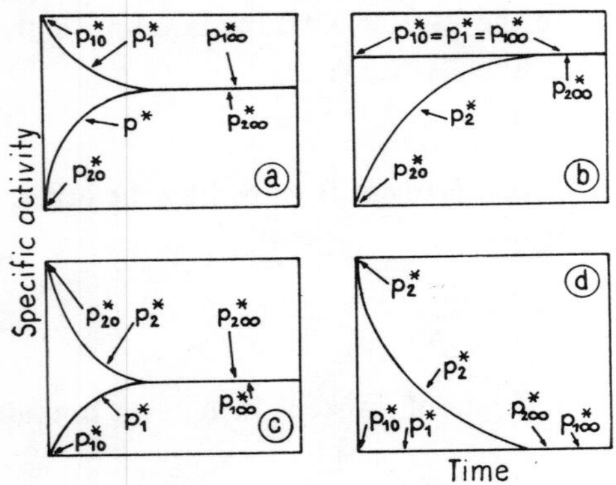

FIGURE 2–2 Time course of the specific activity in wash out (c, d) and uptake (a, b) experiments. Compartment 1: bathing solution; compartment 2: cells.

2.1.2 *Wash out*

This could be a continuation of an experiment of uptake. Once the cells are loaded with tracer we could transfer them to a bathing solution without tracer. They will loose isotope towards the solution. This solution could be a compartment of comparable size (fig. 2–2c) (see for instance Diamond and Solomon, 1959), or may have instead a large amount of substance so as to keep the specific activity of the tracer in this compartment negligible (fig. 2–2d). Aull and Hempling (1963) used this technique to study sodium outfluxes in tumor cells.

To obtain further insight into Compartmental Analysis let us discuss figure 2–3 in which compartment 2 is a thick barrier between compartment 1 and compartment 3, and which consists of two sub-compartments: a very slow one (2a) and a very fast one (2b). If tracer is added to compartment 1, by the time p_3^* reaches a steady value, compartment 2 will also have a quasi-equilibrated specific activity. However, since to obtain the specific activity of the barrier (p_2^*) we have to divide P_2 by S_2 (which includes poorly labeled substance), we will notice that the barrier has achieved an equilibrium at a low specific activity. Compartment 2a is said to constitute a *deep compartment* (sometimes it is also called *hidden* or *masked*). The reader will find

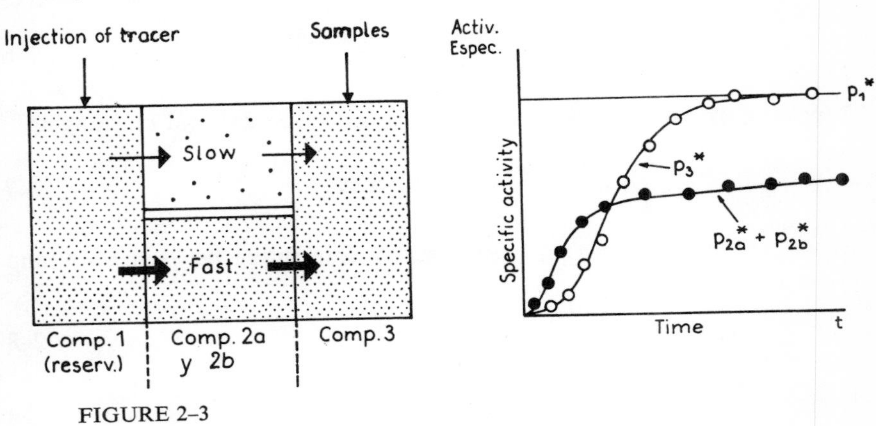

FIGURE 2–3

an illustration of this effect in a work of Katzman and Leiderman (1953) in which the K contained in the brain did not exchange completely with K^{42} in the plasma, and another example in a study by Cereijido and Rotunno (1967) in which they demonstrated that by the time the fluxes across equilibrate, an important fraction of the Na contained in the epithelium remained unlabeled. See also chapter 7.

2.2 *Analysis of the results*

The purpose of this section is to illustrate how to obtain, from the experimental measurements, the values of the fluxes, rate constants and the size* of the compartments.

---

* As defined in page 51 the basic quantity in a compartment is the *amount* of a given substance which it contains. Sometimes the words *size* and *pool* are also used.

Let us analyze the results of a washout experiment of the sort described in page 54 and illustrated in figure 2–2d. The time course of the amount of tracer in the cells (dP_2/dt) is given by

$$\frac{dP_2}{dt} = -J_{21}p_2^* + J_{12}p_1^*. \tag{2–5}$$

Since the system was assumed to be in steady state, the fluxes (J_{21} and J_{12}) are constant. Besides, since the amount of substance in the bathing solution is so large, compartment 1 constitutes a reservoir of negligible specific activity ($p_1^* \simeq 0$). Equation 2–5 then becomes

$$\frac{dP_2}{dt} = -J_{21}p_2^*. \tag{2–6}$$

Replacing the value of p_2^* as described by equation 2–1 and rearranging

$$\frac{dP_2}{P_2} = -\frac{J_{21}}{S_2}\,dt \tag{2–7}$$

where S_2 is also a constant (steady state). Integrating between $t = 0$ and $t = t$ we obtain

$$\ln\frac{P_2}{P_{20}} = -\frac{J_{21}}{S_2}\,t. \tag{2–8}$$

It may be observed that, because of the steady state conditions, if v_2, the volume of the compartment does not vary, the ratio P_2/P_{20} can be substituted by p_2/p_{20}. It could also be substituted by p_2^*/p_{20}^*. Rearranging equation 2–8 we obtain

$$\ln P_2 = -\frac{J_{21}}{S_2}\,t + \ln P_{20}. \tag{2–9}$$

Therefore, if we plot the logarithm of the counts per minute as a function of time as in figure 2–4 we obtain the rate constant k_{21} (see equation 2–2) from the slope and, if we measure the amount of substance in the cell, S_2, we can compute the flux J_{21}.

From the intercept, and using the relationship

$$v_2 = \frac{P_{20}}{p_{20}}$$

we can also measure the volume of compartment 2.

The half life of the substance in the compartment* can also be used as an evaluation of how fast the substance leaves the compartment, in particular when we only want to compare outfluxes in different conditions or among different substances. It is defined as the time ($t_{1/2}$) it takes to reduce any

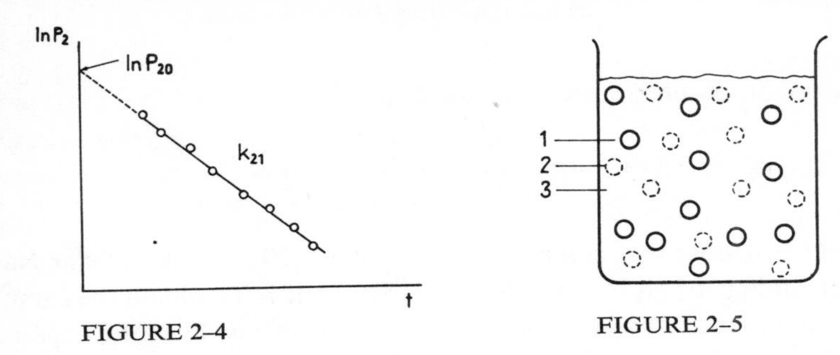

FIGURE 2–4

FIGURE 2–5

value of P_2 to half its value (for instance to reduce P_{20} to $\frac{1}{2}P_{20}$). From equation 2–8 we observe that

$$t_{1/2} = -\frac{S_2}{J_{21}} \ln \frac{1}{2} \quad (J_{ij} = k_{ij}S_i)$$

$$k_{21} = \frac{0.693}{t_{1/2}}. \tag{2–10}$$

It would be convenient to study a further example of the graphical analysis of an experimental curve because it is commonly used in the study of fluxes, even when the tracer is initially in more than one compartment. Figure 2–5 shows to populations of cells (compartments 1 and 2) suspended in an infinite resorvoir (compartment 3). Assume that each population constitute a single compartment in steady state and has a volume of 1 ml. Let us also assume that initially both of them contain Na^{22} with aspecific activity of 10^4 CPM/micromole. The activity in the reservoir remains negligible during the experiment. Na^{22} will be washed out toward the reservoir. The logarithm of P_T/P_{T_0} (the total counts per minute in both types of cells taken as fraction of the initial value) can be plotted *vs.* time as shown in figure 2–6. We observe that in this particular case the wash out consists of a fast and a slow phase, which means that the Na outflux from a type of cells is faster than the outflux from the other. The asymptote of the slow phase has a half life

---

* See requirement *d*, page 52

of 100 hours. Using equation 2–10 we obtain a value of 0.00693 for the rate constant k_{23}. The asymptote has an intercept of 0.3, then the time course of the activity in the slow compartment can be described by the equation

$$\frac{P_2}{P_{T_0}} = 0.3e^{-0.00693t}.$$

Using the condition previously established that

$$p^*_{20} = p^*_{10} = p_{T_0}; \quad \frac{P_{20}}{\text{Na}_2} = \frac{P_{10}}{\text{Na}_1} = \frac{P_{T_0}}{\text{Na}_{\text{TOTAL}}}$$

one can easily deduce that compartment 2 contains 30% of the cellular Na and the remaining 70% is in the cells of compartment 1. To obtain the curve corresponding to the fast compartment we have to substract from the experimental values (say 0.63 at 10 h) the value of the asymptote of the slow phase at the same time (i.e. 0.28). We repeat this operation with all the experimental points and we plot the results. We end up with a new line whose half life is 10 hours, which corresponds to a rate constant k_{13} equal to 0.0693. The intercept, as excepted, is at 0.7. Then the fast compartment is described by

$$\frac{P_1}{P_{T_0}} = 0.7e^{-0.0693t}. \qquad 2\text{–}12$$

The experimental curve is therefore described by the equation

$$\frac{P_{\text{T}}}{P_{T_0}} = 0.7e^{-0.0693t} + 0.3e^{-0.00693t}.$$

Let us assume that the amount of Na contained in the cells is 60 μmole. Hence the Na content in the cells constituting compartment 1 will be

$$S_1 = 0.7 \times 60 = 42\ \mu\text{mole}$$

and the corresponding concentration of sodium will be 42 mM (*i.e.* 42 μmole/ml).

The corresponding values for the other group of cells are 18 μmole and 18 mM respectivelly.

Using equation 2–2 and the values of the S_i and k_{ij} found we obtain

flux from cells type 1 to bathing solution = 2.94 μmole · h^{-1}
flux from cells type 2 to bathing solution = 0.13 μmole · h^{-1}

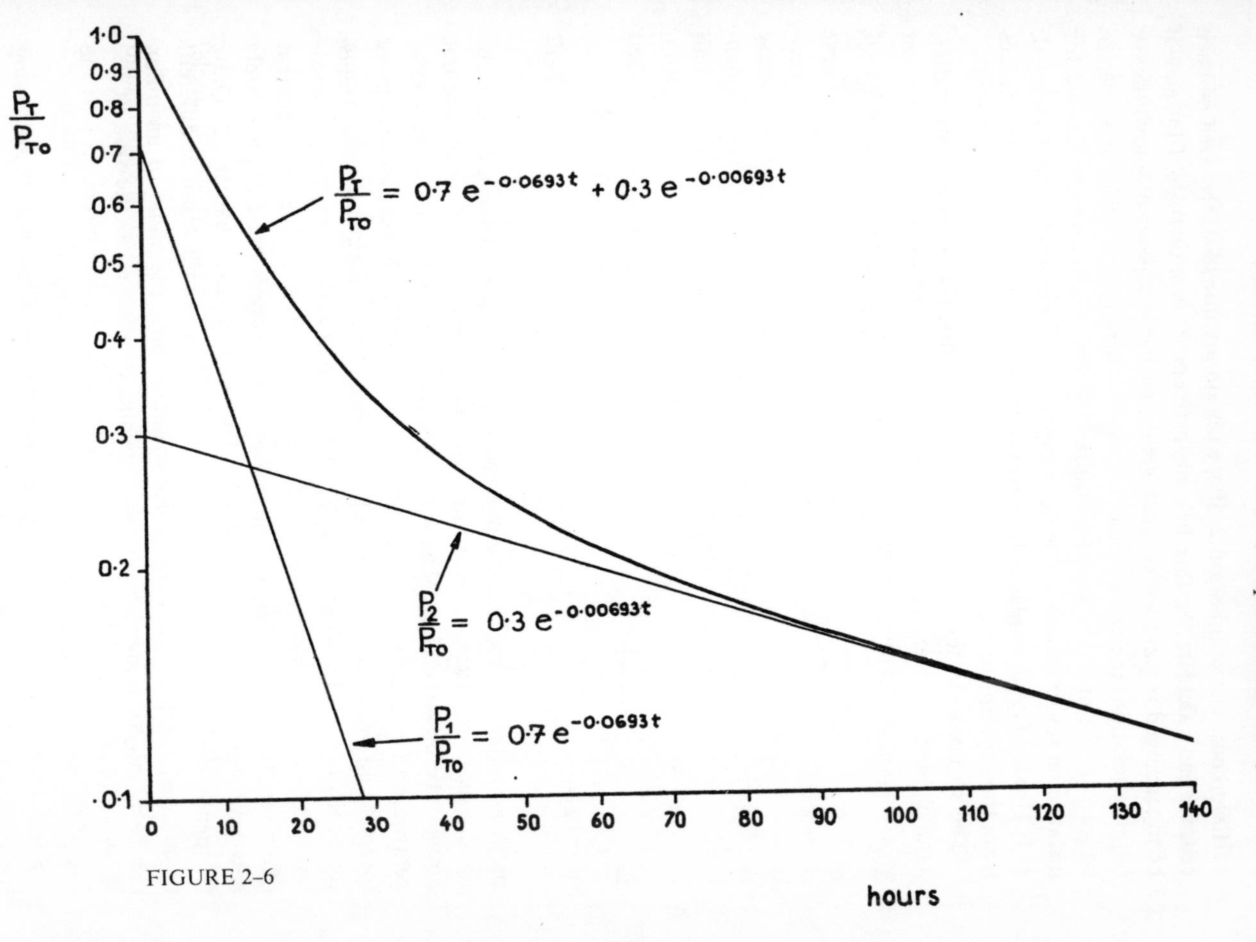

$\frac{P_T}{P_{T0}}$
1·0
0·9
0·8
0·7
0·6
0·5
0·4
0·3
0·2
·0·1
$\frac{P_T}{P_{T0}} = 0{\cdot}7\,e^{-0{\cdot}0693t} + 0{\cdot}3\,e^{-0{\cdot}00693t}$
$\frac{P_2}{P_{T0}} = 0{\cdot}3\,e^{-0{\cdot}00693t}$
$\frac{P_1}{P_{T0}} = 0{\cdot}7\,e^{-0{\cdot}0693t}$
0
10
20
30
40
50
60
70
80
90
100
110
120
130
140
hours

FIGURE 2–6

If we know the number of cells in each population, and the area of each cell, we can calculate the flux in mole h^{-1} cm^{-2}. Notice that, if instead of being arranged in parallel (*i.e.* each group washes independently towards the bathing solution) the two compartments were arranged in series as shown in figure 2–7, the analysis would be quite different. Compartment 1 could for instance represent muscle cells and compartment 2 the intercellular material. A. F. Huxley (1960) explains how to obtain in this case the desired values from the experimental curve.

The variation in the number of compartments, the arrangement (series, parallel, rings), the existence of irreversible steps, the existence of constant or changing reservoirs, the maintenance of steady state, the presence of

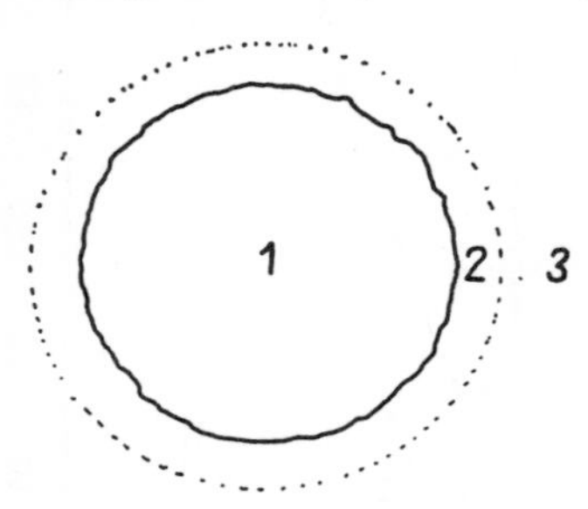

FIGURE 2–7

hidden compartments, and the nature of the parameters that one can technically measure, offers an enormous number of possibilities which are not going to be described here. The reason for this is threefold: 1) As the system gets more and more complicated the measurement of the fluxes becomes more indirect and less accurate. Even when we can derive an equation describing the behaviour of the system, the physical meaning and the error in the evaluation of each parameter may rend the analysis useless. 2) The measurement of a given flux in a particular biological system generally presents an entirely new problem. Some examples will be described in later chapters. 3) Quite frequently the evaluation of a flux is a result of the combination of different techniques. Chapter 7 illustrates, for instance, how the value of an influx can be obtained, under proper experimental conditions, as the sum of the electrical current carried by the net flux of an ion in one direction plus the tracer flux of the ions measured in the opposite direction.

Some time the analysis of a system benefits from the use of an analog computer in which the amount of substance in the different compartments

(S_1, S_2 ··· S_n) is represented by the capacity of adjustable capacitances. The capacitances are connected by conductors according to the arrangement of the compartments in the biological system which is being studied. Each of these connections includes a variable resistance and represents a flux (J_{12}, J_{21} ··· J_{ij}). If, because of previous measurements, we know the values of some fluxes or the size of a compartment we set the values in the computer. The "tracer" is simulated by a charge introduced into one of the capacitances. The behaviour of a given compartment can be followed by observing on the screen the variations of voltage (which represents specific activity) versus time. The trace of the cathode ray can be varied by adjusting the resistances

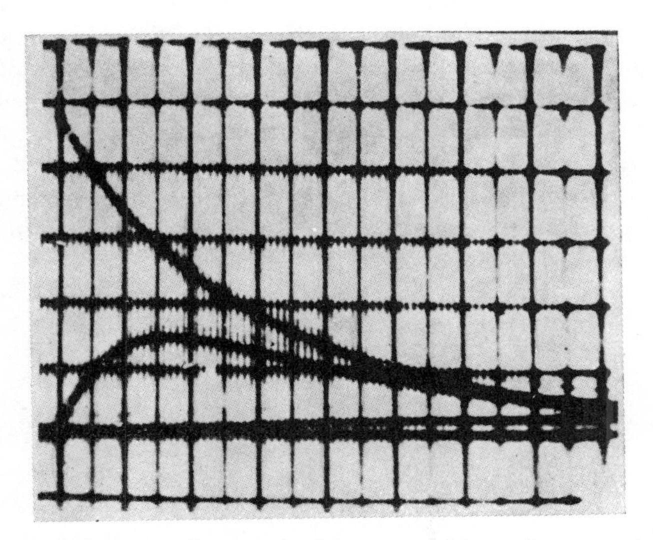

FIGURE 2–8 Photograph of the screen of the analoge computer cathode ray tube. Ordinate: specific activity of K^{42} in the plasma (top curve) and in two intracellular compartments. Abscissa: time. Each small divition is equal to 30 minutes (Taken with kind permission from A.K. Solomon, 1960).

(fluxes) until the curve fits the experimental points. The answer is given in relative values (for instance $J_{12} = 3.5J_{23}$). Absolute values of the fluxes can be deduced when the information on the real value of the parameters involved, and which where initially fed into the computer, is enough. Even if we do not have enough information as to get absolute values we could learn for instance that a given flux cannot be larger than some other flux, or will pinpoint the information we have to seek in further experiments to obtain the required values.

Notice that this explanation assumes that we have a great deal of information on the system, such as size, number and connections of the compartments, etc. Figure 2–8 illustrates a study of the distribution of K^{42} between the plasma (top curve) and the intracellular compartments carried out by A.K. Solomon (1960). The ordinate represents the specific activity in an arbitrary scale. The abscissa represents time. A detailed description of the uses of analog computers in Compartmental Analysis is given by Sheppard (1948). As pointed out by A.K. Solomon (1960), the computer is a very useful tool when the specific activity in one of the compartments we are studying varies in a complicated way. In this case, instead of solving the equation for this compartment we can simulate the time course of its specific activity, and use it as the imput function for the next compartment.

As we said at the beginning of this chapter, one of the roles of membranes is to act as barriers between compartments. Now we have a valuable tool to study how substances, under given conditions get across the membrane. In following chapters we will study how it is used to obtain information on the nature of biological membranes.

3

Theoretical approaches to the flux of substances across biological membranes

Now that we have an idea of the structure of the membrane, and have learned some techniques to measure the fluxes and distribution of substances across it, we are going to start the description of how substances get in or out of the cells and how they achieve a particular steady state distribution. The mere enumeration of some of the phenomena that this type of studies has to elucidate is impressive: 1) Some substances in the cells achieve a steady state concentration which is much higher (*e.g.*: potassium) or much lower (*e.g.*: sodium) than the concentration in the bathing medium. 2) Some substances penetrate the cells with a speed which is proportional to their solubility in lipids (see in figure 1–1 erythritol, dicyandiamide, monacetin, antipyrin) while others do it much faster than their lipid solubility would predict (Na^+, K^+, glucose). 3) While the passage of some substances is inversely proportional to the radius of their molecules; the passage of other substances (like L-arabinose and D-arabinose) which have the same radius might be different by orders of magnitude. 4) The fluxes of some substances increase linearly with the difference in concentration between both sides of the membrane; the flux of some others rise to a point and then stay there in spite of further increases of the concentration gradient. 5) the flux of a substance (for instance glucose across the intestinal mucosa) is competed by the passage of another substance (for instance by galactose) but might be helped by the flux of some others (for instance by Na^+). 6) Some substances entering a cell provoke the exit of substances present in the cytoplasm; some others prevent or impede it. 7) The passage and distribution of some substance depends on the supply of metabolic energy. 8) A membrane may be a hundred times more permeable to K^+ than to Na^+, suddenly became more permeable to Na^+ than to K^+ and then recover the initial state, all in a few milliseconds. 9) A membrane may reversible change its permeability in response to the presence of a hormone.

The general approach to these problems might be summarized as follows: one tries to explain fluxes and distribution on the basis of the driving forces present, using theoretical approaches suitable to the conditions and assumptions that the system permits. When the relationship between fluxes and forces, or the final distribution achieved, is not what the theory would predict, one attributes the effect to some special entity in the membrane. Thus the cell membrane, born almost a century ago as a simple semipermeable barrier, today is thought to possess charged and uncharged pores, mono- or divalent carriers, dimerizers, fixed sites, gates, neutral or electronic pumps, etc. We will study the flux of substances following the same pattern: we will first review the relationship between fluxes and forces according to three of the theoretical approaches generally used. Next we will analyze those fluxes of substances that have forced to assume the existence of some of the mechanisms mentioned above and we will illustrate also how they work (chapter 4). We will leave for chapter 5 the study of fluxes associated to the metabolism and for chapter 6 the movement and distribution of electrolytes.

A wide variety of theoretical approaches have been used. Classification of these theories are intended mainly to compare the advantages of the different approaches (Schlögl, 1956; Laksminarayanaiah, 1965). They can be divided into three main groups: 1) Approaches based in the Nernst-Planck flux equation, 2) approaches based in the Thermodynamics of Irreversible Processes and 3) approaches based in the Theory of Absolute Reaction Rates.

3.1 Approaches based on Nernst-Planck flux equation

The flux per unit area may be expressed as (Teorell, 1953)

$$\text{Flux} = \text{mobility} \times \text{concentration} \times \text{force}. \qquad 3\text{–}1$$

The mobility used here is ω, the molar mobility per unit force (cm. mol. sec^{-1} $dyne^{-1}$). When dealing with electric currents (chapter 6) we shall use a somewhat different mobility: u (cm^2 sec^{-1} $volt^{-1}$). One can adapt this treatment to study biological membranes. According to the ideas

of Nernst and Planck the driving force is the algebraic sum of all the forces acting upon the membrane. The forces operating across biological membranes of the sort discussed in chapters 4 to 6 are proportional to*:
1) The chemical potential gradient $d\mu/d_x$. Since in most treatments the cytoplasmal solutes are assumed to be contained in free solution and their concentration is generally below 0.2 M, the activity coefficient is assumed to be 1 and the concentration gradient dc/dx is used in the calculation of $d\mu/dx$. Until quite recently the techniques for direct measurement of the chemical activities inside the cells were not available and concentrations, instead of activities, had to be used anyway. In chapter 9 we consider a different view in which a departure from a state of free solution plays a fundamental role. 2) The electrical potential gradient $d\psi/dx$. 3) The pressure gradient dP/dx. The total force resulting is represented by the negative gradient of electro-chemical potential $d\bar{\mu}/dx$.

$$-\frac{d\bar{\mu}}{dx} = -RT\left[\frac{1}{c}\frac{dc}{dx} + \frac{\bar{v}}{RT}\frac{dP}{dx} + \frac{zF}{RT}\frac{d\psi}{dx}\right]. \qquad 3\text{–}2$$

Introducing this equation into equation 3–1 we obtain

$$J = -\omega cRT\left[\frac{1}{c}\frac{dc}{dx} + \frac{\bar{v}}{RT}\frac{dP}{dx} + \frac{zF}{RT}\frac{d\psi}{dx}\right]. \qquad 3\text{–}3$$

To use this equation across a given membrane one has to integrate it first. This requires a considerable amount of information on the structure and thermodynamic properties of the membrane. Integrations carried out by different authors under a wider variety of boundary conditions have played a major role in the study of kinetic and steady state properties of the membranes. We will use this equation in particular when dealing with electrical potentials and ion permeation in chapter 6. For typical examples of the use of this approach read Teorell (1953); Schlögl (1954); Mackie and Meares (1955); Helfferich (1956, 1962); and Conti and Eisenman (1965).

Now we are going to make a drastic simplification of equation 3–3 to illustrate some of the most common difficulties that biological membranes present and the typical assumptions one makes to overcome them.

---

* Fluxes considered in this book are normal to the membrane surface (x-coordinate) and are constant in steady state. Absolute values and differential quotients (d/d_x) are used as gradients.

When the diffusing molecule does not have a net charge ($z = 0$) and there is no difference in hydrostatic pressure ($dP/dx = 0$). Equation 3–3 reduces to

$$J = -\omega RT \frac{dc}{dx}. \qquad 3\text{–}4$$

It can be shown (Einstein, 1905) that the diffusion coefficient D is equal to

$$D = \omega RT. \qquad 3\text{–}5$$

Introducing D into equation 3–4 it gives

$$J = -D \frac{dc}{dx}. \qquad 3\text{–}6$$

We usually do not know all the necessary boundary conditions to integrate this equation so as to obtain a useful equation. Therefore, we are forced to make further assumptions. Typical assumptions in the study of biological membranes are: 1) That the concentration gradient across the membrane is homogeneous. 2) That the diffusion path is straight and perpendicular to the plane of the membrane so that its length (l) is equal to the thickness of the membrane. Integrating 3–6 under these conditions we obtain

$$J = -D \frac{c_1' - c_2'}{l}. \qquad 3\text{–}7$$

But even to use this equation we need some information that we usually do not have (see figure 3–1). One thing we ignore is l, the thickness of the membrane. Another one is D, the diffusion coefficient of the substance *within* the membrane. We dot not know the concentrations c_1' and c_2' either. We could assume that the concentrations in the membrane phases which are immediately in contact with the solutions, are related to the concentrations in the solutions through a partition coefficient as follows

$$c_1' = \beta c_1 \quad \text{and} \quad c_2' = \beta c_2 \qquad 3\text{–}8$$

so that equation 3–7 can be transformed to

$$J = -\frac{D\beta}{l}(c_1 - c_2)$$

$$J = P(c_1 - c_2) \quad \text{where} \quad P \equiv -\frac{D\beta}{l}. \qquad 3\text{–}9$$

So we started out with a general formula which allowed us to take into account most of the forces operating to move a given substance through a membrane and ended up with Fick's law which describes one of the simple ways of flowing: by simple diffusion. When the diffusing substance does have a net charge, or a pressure is present or both, the situation is more complicated and the membrane may show electrical phenomena such as diffusion potentials, streaming potentials, etc.

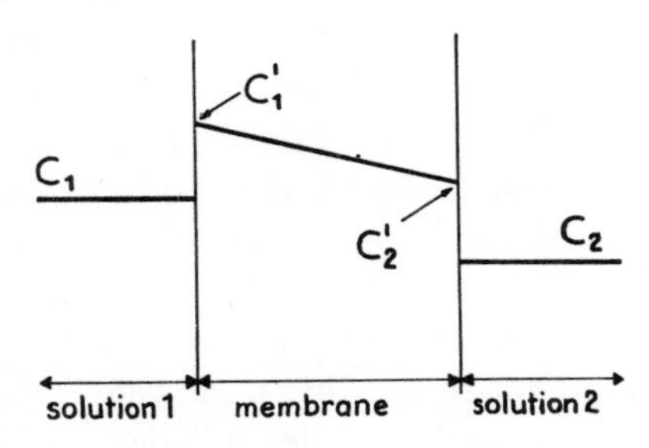

FIGURE 3–1

There are two points woth noticing: 1) The flux described by equation 3–3 is a *net* flux, *i.e.* the difference between two unidirectional fluxes (J_{12} and J_{21}) going in opposite directions through the same membrane. They may interact giving rise to peculiar phenomena some of which will be discussed in later chapters. 2) When more than one species are involved, we need a whole set of Nernst-Planck equations. Living membranes are continuously penetrated by a large number of substances at the same time and in both directions. The electrical potential generated by the diffusion of a given species may accelerate or restrict the penetration of another ion species. The Nernst-Planck equation is powerful in describing artificial systems with one substance moving at a time.

3.2 *The approach based on thermodynamics of irreversible processes*

Two groups of considerations underly this approach: 1) The use of thermodynamic concepts specially developed to deal not just with equilibrium, but with steady state. 2) The view that all the fluxes in the system may be related to all the driving forces present. Therefore, we shall elaborate on these two points.

3.2.1 *Equilibrium vs non-equilibrium thermodynamics*

The thermodynamics that is often called "classic" deals mainly with closed-static systems in equilibrium, in which no spontaneous process takes place and whose properties are not time dependent. It is, therefore, inadequate to use it to study systems with net fluxes of matter, heat or electricity through their boundaries. Even when the biological systems with net fluxes across their boundaries could also achieve a time-independent state and all their parameters become nearly-constant, they are not in equilibrium: they are in *steady state*. A bacteria, for instance, could maintain a constant concentration of glucose thanks to a perfect balance between the net influx through the membrane and the consumption of glucose by the metabolism. A given cell may have a high and constant concentration of potassium (as compared with the bathing solution) due to its impermeability to K^+. Another kind of cell may also maintain a high and constant concentration of K^+ (not explained by a negative electrical potential nor by negatively charged macromolecules inside the cytoplasm) through a perfectly balanced influx and efflux of K^+ across the membrane. The basic difference between the two cells is that the second one maintains its concentration of K^+ through a *process* (a one than in chapter 5 we will call active transport) which, therefore, produces entropy. The higher the permeability to K^+, the larger the leak of K^+ and the larger the amount of K^+ that the cellular membrane has to pump back to the cytoplasm to maintain constant the cellular concentration of K^+. This example shows the interest in obtaining a formal relationship between the rate of entropy production and the fluxes in systems which are not at equilibrium but not too far from it either. The *first* step is to take the total entropy change (dS), and discriminate the change due to internal production by the system (d_iS) from the entropy changed as a consequence of the interaction with the environment (d_eS) (Guggenheim, 1950)

$$dS = d_eS + d_iS \tag{3–10}$$

d_eS can be expressed as a function of the heat absorbed by the system (dq)

$$d_eS = \frac{dq}{T} \tag{3–11}$$

so that equation 3–10 can be transformed into

$$d_iS = dS - \frac{dq}{T}. \tag{3–12}$$

The First Principle tells us that, for a system doing only expansion work the following relation should hold

$$dq = dU + P\,dV \qquad 3\text{–}13$$

where dU is the change in internal energy and dV the change in volume. Introducing equation 3–13 into 3–12 we obtain

$$d_iS = \frac{T\,dS - (dU + P\,dV)}{T}. \qquad 3\text{–}14$$

If the system is not too far from equilibrium, we may use the Gibb's free energy defined as

$$G = U + PV - TS$$

and restrict our treatment to systems at constant P and T. So that equation 3–14 becomes

$$d_iS = -\frac{dG}{T}.$$

The *second* step is the introduction of *time*, so that the *rate* of entropy production will be proportional to the *rate* of change of the Gibb's free energy

$$\frac{d_iS}{dt} = -\frac{1}{T}\frac{dG}{dt}. \qquad 3\text{–}15$$

To illustrate this point let us consider the change in entropy production and free energy in a chemical reaction

$$(dG)_{TP} = \sum_i \mu_i\,dn_i$$

$$\frac{d_iS}{dt} = -\sum_i \frac{\mu_i}{T}\frac{dn_i}{dt}. \qquad 3\text{–}16$$

Defining the degree of advancement of the chemical reaction as (de Donder and van Rysselberghe 1936)

$$d\xi = \frac{dn_i}{\nu_i}$$

where ν is a stoichiometric coefficient. Introducing this parameter into equation 3–16 we obtain

$$\frac{d_iS}{dt} = -\sum_i \frac{\mu_i\nu_i}{T}\frac{d\xi}{dt}$$

de Donder calls "affinity" (A) of the reaction the term $-\sum_i \mu_i \nu_i$ which is also equal to the ΔF of the reaction (Lewis and Randall, 1961)

$$T \frac{d_i S}{dt} = A \frac{d\xi}{dt}.$$

This equation tells us that the production of entropy is a function of the reaction rate ($d\xi/dt$) times a "thermodynamic force" (A). The term of the left is what Lord Rayleigh (1873) called "dissipation function" (Φ). The whole equation is of the form

$$\Phi = JX$$

that is, the "dissipation function" is equal to the product of a flux times a force (X). It can be shown that it is a particular case of the more general equation

$$T \frac{d_i S}{dt} = \sum_i J_i X_i. \qquad 3\text{–}17$$

3.2.2 The relationship between fluxes and forces

The second characteristic of the application of Non-equilibrium Thermodynamics to the study of membranes is the assumption that each flow depends on all the driving forces present in the system and not merely the one "classically" associated (herefrom called *conjugated force*). Classically each flux J_i is originated by its conjugated force X_i. Electrical flow, for instance, depends on the electrical potential difference; the flow of heat on the temperature gradient; the flux of molecules on the chemical gradient and so on. The new approach, instead, assumes that if the n fluxes present in a system are sufficiently slow, they are related to *all* the driving forces in the following manner (Onsager, 1931a and b).

$$\begin{aligned} J_1 &= L_{11}X_1 + L_{12}X_2 + \cdots + L_{1n}X_n \\ J_2 &= L_{21}X_1 + L_{22}X_2 + \cdots + L_{2n}X_n \\ &\cdots\cdots\cdots\cdots\cdots\cdots\cdots\cdots \\ J_n &= L_{n1}X_1 + L_{n2}X_2 + \cdots + L_{nn}X_n \end{aligned} \qquad 3\text{–}18$$

writing the same in a more condensed form

$$J_i = \sum_{j=i}^{n} L_{ij}X_j. \qquad 3\text{–}19$$

Familiar examples are found in thermoelectricity where a gradient of electrical potential originates not only a current but also a flow of heat and, conversely, a temperature gradient can give rise to an electrical current. On the basis of these considerations the flux J_i of a substance can be considered as the sum of a flux due to its own (conjugated) driving force and a flux arising from the operation of other driving forces present. So

$$J_i = L_{ii}\Delta\bar{\mu}_i + \sum_{j \neq i} L_{ij}\Delta\bar{\mu}_j. \qquad 3\text{–}20$$

The coefficient L_{ii} associates the flux J_i with its conjugated force X_i (in this case X_i is $\Delta\mu_i$, the electrochemical potential gradient). The coefficients L_{ij} are the *coupling coefficients* which determine how much a non-conjugated driving force (which is the conjugated force of some other flux) will influence the flux J_i. The term

$$\sum_{j \neq i} L_{ij}\Delta\bar{\mu}_j$$

in equation 3–20 represents the contribution that the forces conjugated to other fluxes make to the flux J_i. Of course the main contribution to a given flux generally comes from the operation of its own conjugated force. The point is that a given flux may occur in the absence of its "typical" driving force or even go in the opposite direction (a very common situation in biological membranes) simply because the component of the flux due to non-conjugated forces balances the one due to the conjugated force.

If all the coupling coefficients (L_{ij}) were zero, the flux J_i would be driven only by its conjugated force and equation 3–20 would reduce to a "classical" law. For instance, let us use the electrochemical potential gradient between substance i in compartment 1 and compartment 2 as

$$\Delta\bar{\mu}_i = RT \ln \frac{(c_i)_1}{(c_i)_2} + zF\Delta\psi. \qquad 3\text{–}21$$

Now, assume that the substance i is not charged ($z = 0$) and introduce the value of $\Delta\bar{\mu}_i$ into equation 3–20

$$J_i = L_{ii}\, RT \ln \frac{(c_i)_1}{(c_i)_2} \qquad 3\text{–}22$$

using the series

$$\ln x = 2\left(\frac{x-1}{x+1}\right) + \frac{1}{3}\left(\frac{x-1}{x+1}\right)^3 \cdots$$

and making $x = c_1/c_2$, we may transform the logarithmic term in equation 3–22. If the system is not far from equilibrium c_1/c_2 is close to unity and we can take only the first term of the series. Equation 3–22 then becomes

$$J_i = 2L_{ii}RT \frac{c_1/c_2 - 1}{c_1/c_2 + 1}. \qquad 3\text{–}23$$

Defining permeability as

$$P \equiv \frac{2L_{ii}RT}{c_1 + c_2}.$$

Equation 3–23 becomes

$$J_i = P(c_1 - c_2).$$

In other words, if there is no interaction between the different fluxes, and the system is close to equilibrium we obtain an expression equivalent to Fick's law.

The advantage of the approach based on Non-equilibrium Thermodynamics is that it does not require a detailed knowledge of the structure and thermodynamic properties of the membrane. It is also a suitable theory for systems such as cellular membranes, which have a larger number of fluxes going on at the same time and in both directions, and which are known to exhibit phenomena due to interactions between fluxes.

During the last few years a large effort was made in devising criteria of stability and active transport and to study interactions between solvent, solutes and membranes (Kedem and Katchalsky, 1958, 1961; Kedem, 1960; Katchalsky and Kedem, 1962; Dainty and Ginzburg, 1963; Kimizuka and Koketsu, 1964; Marro and Pesente, 1964a and b; Katchalsky and Curran, 1965; Katchalsky and Spangler, 1968). One of the main difficulties is that this approach is strong where fluxes are slow, and the relationship with the forces is linear, a condition seldom found in biological membranes.

For further information on the Thermodynamics of Irreversible processes see: Denbig (1951), de Groot (1951), Prigogine (1960) and Kirkwood and Oppenheim (1961).

3.3 *Approaches based on the absolute reaction rate theory*

This approach is powerful in crystals in which we already know the molecular organization and we want to find out which particular mechanism—among several possible—is operating the migration of a given molecule. As we have seen in chapter 1, biological membranes have a high (liquid-crystal) degree of organization. The water solutions in contact with the membranes could also be considered to possess a quasi-crystalline organization (Glasstone, Laidler and Eyring, 1941; Frenkel, 1946; Kavanau, 1965). The diffusion of substances in the solution-membrane-solution sequence can be therefore compared to the migration of particles in solids, the main energy jumps being at the solution-membrane boundaries.

The architecture of the cell membrane is too complicated, and our knowledge not detailed enough, to take full advantage of this approach. However, there are some well-known results of other related fields, which can be applied to membranes in a straightforward manner. For instance, if we observe that the passage of a substance through a biological membrane depends on the temperature and/or concentration in a manner which, in solids or ion exchangers, is known to be due to a particular migration mechanism (*e.g.*: a vacancy mechanism), we may infer that the substance crosses the membrane through the same mechanism. In studying ion migration (see chapter 6) we will realize that some mechanisms, that were originally proposed for ionic crystals, give the only consistant model to explain some current-voltage relationship and the diffusion-mobilities discrepancies that we observe in biological membranes. Besides, this method would become more and more useful as our knowledge of the membrane structure improves. So it would be convenient to make a description of some of the basic ideas in diffusion in solids and ionic crystals and then discuss briefly its possible application to biological membranes.

When two atoms approach each other, repulsive forces push them apart. When they separate repulsion forces become too weak and the atoms are attracted. When they are at a certain distance, repulsive forces balance the attractive forces and the atoms remain at equilibrium. In a crystal an atom maintains the same respulsive-attractive relationship with its neighbours and it is confined to vibrate thermically in a restricted place. Each time it tries to escape it is repelled back by its neighbours.

Not only the atoms forming the lattice of the crystal find it hard to escape. This also holds for atoms in the interstice of the lattice. The atom represented in black in figure 3–2 for instance would have to overcome a strong repulsion from its neighbours to move from its present position to the interstice on the right across the hatched barrier. Ocassionally a thermal fluctuation of the atom would be strong enough to make it jump. In order to jump the atom needs an activation energy E^*. From the new position the atom could (when another large thermal fluctuation comes) jump back to its previous position or to a position further right, or up, etc. Statistical mechanics tells us that the frequency of successful jumps Γ is given by

$$\Gamma = \nu e^{-E^*/kT} \qquad 3\text{–}24$$

where ν is the atomic vibration frequency, and k is the Boltzmann constant.

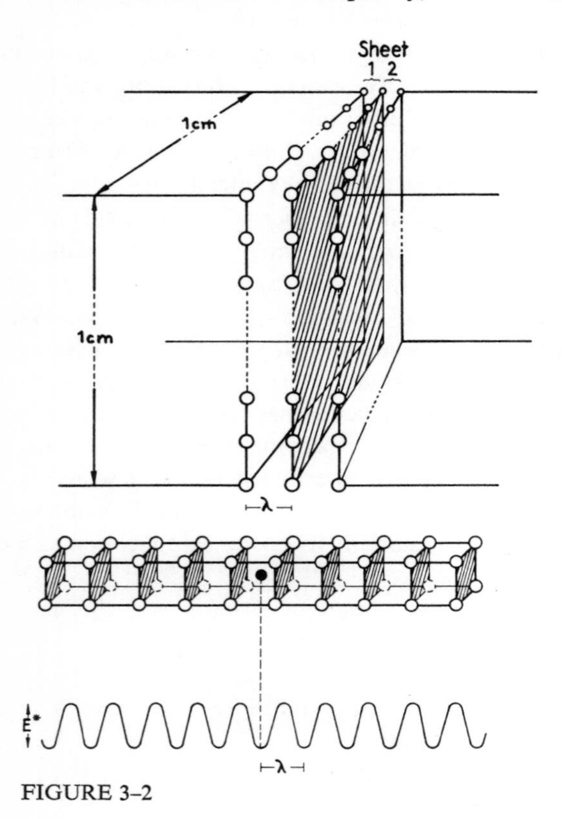

FIGURE 3–2

Consider we have n_1 impurity atoms in a sheet of crystal of a thickness equal to λ and of an area of 1 cm^2 (sheet 1 of figure 3–2). The concentration of impurities in this sheet will be

$$c_1 = \frac{\eta_1}{\lambda}.$$

The impurities will jump out of their position with a frequency Γ. Even assuming that no other factors prevented them from going to the right, only a fraction θ of them will jump in this particular direction. So that J_{12} (the flux from sheet 1, sheet 2) will be

$$J_{12} = \Gamma\theta n_1$$

and replacing n_1 by λc_1 we obtain

$$J_{12} = \Gamma\theta\lambda c_1 . \qquad 3\text{–}25$$

The flux in the opposite direction (sheet 2 to sheet 1) will follow a similar behaviour, so that the net flux would be

$$J_{\text{net}} = \Gamma\theta\lambda\,(C_1 - C_2) \qquad 3\text{–}26$$

since λ is very small (a few Ångstroms) we may take

$$\frac{C_1 - C_2}{\lambda} = -\frac{dC}{dx} .$$

Equation 3–26 then becomes

$$J = -\Gamma\theta\lambda^2 \frac{dC}{dx}$$

$$J = -D\frac{dC}{dx} \qquad 3\text{–}27$$

which is Fick's First law of diffusion (see equation 3–6) where

$$D = \Gamma\theta\lambda^2 . \qquad 3\text{–}28$$

This simple case describes the diffusion of an interstitial impurity.

If instead of being interstitial the atom whose migration we are studying occupies a place in the lattice and migrates through a vacancy mechanism, the situation is somewhat different, but can be handled along the same line. This time, though, an atom not only requires certain activation energy E^* to migrate but also needs a vacancy (a lattice point where the corresponding atom is missing) where to jump.

When all atoms are vibrating in their place in the crystal the Minimum Energy Principle is satisfied. However the Maximum Entropy Principle is not, and the system will tend to less organized states in which atoms will escape from their positions leaving vacant sites. The two principles compromise and a situation is reached where there is a fraction of vacant sites (f_v). If E_v is the energy to form a vacancy and S_v is the entropy gained when the vacancy is formed, the fraction of vacant sites in a pure crystal will be

$$f_v = e^{S_v/k}\, e^{-E_v/kT} \qquad 3\text{–}29$$

so that if the simple vacancy mechanism is involved, the diffusion coefficient is not the one given by equation 3–28 but

$$D = \Gamma\theta\lambda^2 f_v \qquad 3\text{–}30$$

likewise we could find diffusion coefficients for other migration processes. The interested reader can consult: Barrer 1941; Yost 1952; Dekker, 1957; Lidiard, 1957; Shewman, 1963.

Taking the value of Γ from equation 3–24 and the value of f_v from equation 3–29 we obtain

$$D = \lambda^2\theta\nu\, e^{-E^*/kT}\, e^{S_v/k}\, e^{-E_v/kT}. \qquad 3\text{–}31$$

Inspecting equation 3–31 we notice that the diffusion coefficient consists essentially of two factors: one that is a constant and one that is an exponential function of both, temperature and energy. Vacancies are a particular kind of point defect on which diffusion can be based, but these considerations apply to other migration processes as well. The energy term depends on the sort of mechanism involved in the migration and it is often symbolized by a Q. These considerations permit to write all equations for the diffusion coefficient of the different mechanisms as

$$D = D_0\, e^{-Q/kT} \qquad 3\text{–}32$$

which is called the Arrhenius equation for the diffusion coefficient. In the particular case of equation 3–31 Q and D_0 are

$$Q = E_v + E_v^* \qquad D_0 = \lambda^2\theta\nu\, e^{S_v/k}.$$

Taking logarithms, equation 3–32 becomes

$$\ln D = -\frac{Q}{k}\frac{1}{T} + \ln D_0.$$

By studying diffusion at several temperatures one obtains a set of values of the diffusion coefficient that, ploting the logarithm of D as a function of $1/T$ gives the heat of activation Q from the slope, and D_0 from the intercept. As said above, different migration mechanisms have different energies involved, this is why this sort of study gives insight into the mechanism of migration. Elucidation of the mechanism involved in migration would be the ultimate aim of permeability studies, but the equations used in absolute rate treatment ask for information on the structure we do not posses yet.

Following a pioneer series of studies carried out by Danielli (see Davson and Danielli, 1943), efforts were made to adapt the absolute rate theory treatment of diffusion to the study of cell membranes. Thus Zwolinski, Eyring and Reese (1949) constructed an energy profile of the type depicted in figure 3–3 representing the transport of matter through a solution-membrane-solution system. First they took into consideration the concentration of particles at various minima under steady state conditions, similarly to the concentration in a given plane sheat that we have illustrated in

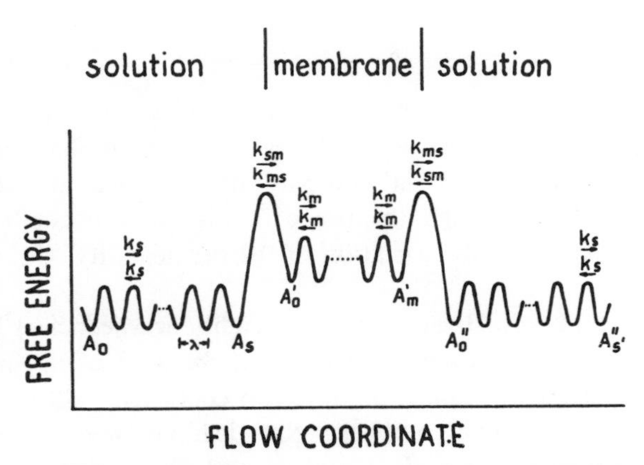

FIGURE 3–3 Energy profile curve for the transport of matter through a solution-membrane-solution system. (After B.J. Zwolinski, H. Eyring and C.E. Resse, 1949)

figure 3–2. They also defined rate constants for the forward and reverse flow. They reached an expression in which the movement of matter is governed by the relative heights of the potential barriers (the one the migrating atom has to overcome in going from one position to the other). Next they incorporated the external forces acting on the kinetic system. The basic assumption is that an external force acting on a single unit process (a jump) provides an amount of work which tends to aid or hinder the process by increasing or decreasing the free energy of the initial and final positions. Once they got equations relating fluxes and forces they went on to illustrate their use in actual systems. One of them refers to membranes and concerns us here. It is precisely the system depicted in figure 3–3. Zwolinski *et al.* have shown that the permeability of this system is a function of the distance between equilibrium maxima (λ), the number of jumps on the solutions on both sides (s and s') and in the membrane (m) and of the rate constants defines as follows:

k_s: constant for diffusion in solution.
k_m: constant for diffusion in the membrane.
k_{sm}: constant for diffusion through solution-membrane interface.
k_{ms}: constant for diffusion through membrane-solution interface.

$$P = \frac{k_s \lambda}{s + s' + 2\dfrac{k_s}{k_m} + m\dfrac{k_s k_{ms}}{k_{sm} k_m}}. \qquad 3\text{–}33$$

By making simplifying assumptions Zwolinsky *et al.* have also shown the application of equation 3–33 to the permeability of water, aliphatic alcohols, ethers and amides through animal and plant cells. They analyzed the mechanisms of permeation through the use of a distribution coefficient (given by the ratio k_{sm}/k_{ms}) and the variation of the permeability with temperature.

The probability of finding a divacancy (two neighbour vacancies) is of course smaller than finding a vacancy. Therefore if because of its size, a migrating molecule needs a divacancy, its diffusion coefficient will be low. This intuitively shows that there is an inverse relationship between molecule size and diffusion coefficient. Formal discussion and illustrations of this point can be found in Davson and Danielli (1943) and Stein (1967). When the size of the diffusing molecule is very big as compared to the size of the solute (fig. 3–4) a situation is reached in which it is easier for a solvent molecule to find a place "behind" the diffusing molecule of solute, than for the diffusing solute to find a whole row of vacancies afore of it. Thus, in this case, the solute diffuses largely by diffusion of the solvent molecules in the opposite direction (Stein, 1962).

So far we avoided the discussion of charged species so as to simplify the description of the basic approaches. Ion migration will be the subject of chapter 6. However it would be adequate to point out here that the Absolute Rate approach, as applied to ionic crystals, semiconductors and ion-exchangers gives valuable information on the process of ion movement which is later extrapolated to biological membranes. One of the most fertile sources of information on the intrinsic processes of ion permeation has been the study of non-living membranes made out of glasses, polyelectrolytes, metals, lipids, etc. Typical outcomes of these studies are the suggestion that migration through membranes involvs Frankel defects (Ilani, 1966) obeys Elovich's equation (Cope, 1965), etc. Considering that these phenomena are characteristic of ionic crystals, glasses and metals, and that they are usually studied using the absolute rate approach, it follows that this approach would greatly benefit the understanding of biological membranes. A description of diffusion in metals and ionic crystals would be very useful but, even

FIGURE 3–4 Diffusion of a large molecule on a lattice model (Taken with kind permission from W.D. Stein, 1962a).

a somere one, would take too long a digretion as to justify its inclusion here. We only wanted to sketch the characteristics of one of the many approaches: the one based on the concept of absolute rates. Its attraction stems from its fundamental character. Its difficulties arise from the requirement of a detailed knowledge of membrane structure. Another difficulty, which concerns biologists in particular is the following: one of the most important tools in the study of migration is the observation of the temperature dependence of the diffusion coefficient and, in particular, the use of the Arrhenius plot. In non-biological systems one may explore diffusion in a wide range of temperatures (several hundreds of degrees). In dealing with biological systems, we are confined to a range of a few degrees in which the preparation would not be damaged.

4

Fluxes II—The role of the membrane

IN THE first chapter we have described the architecture of the biological membrane. In the second we have learned some techniques to measure how things move and distribute in membrane systems. In the third we have described some of the theories relating forces and fluxes across membranes. Now that we have some techniques and a set of theoretical approaches we shall devote three chapters (4 to 6) describing how things get across biological membranes. In this chapter we will analize those factors which cannot be ascribed to the driving forces alone, and have to be accounted for by special properties of the membrane.

4.1 Simple diffusion

If the diffusion coefficient of a substance through the membrane is not zero, a net amount of substance will cross the membrane whenever its electrochemical potential is higher on one side of the membrane than on the other. If in doing so it obeys equation 3–9, we would normally maintain that it enters by simple diffusion. Simple diffusion can be generally observed in two cases: a) when the unique mechanism of penetration is by simple diffusion. This is illustrated in figure 4–1, taken from a paper by Cohen and Monod (1957) in which they studied the influx of a β galactoside into bacterial cells of the "cryptical" type. The influx is linearly related to the concentration of galactoside in the bathing medium.
b) when a substance uses more than one mechanism of permeation. If the concentration of the substance is high enough, many of these mechanisms reach a maximal rate of operation; thereon an increase of the concentration does not bring about an appreciable rise in the part of the influx carried by

that mechanism, and we say that the mechanism is *saturated.* On the contrary, as implicit in equation 3–9, simple diffusion does not saturate. Therefore, simple diffusion is often detected at high concentrations, *i.e.* once the other mechanism become saturated. Figure 4–2 shows an experiment in which Reisin and Cereijido (1969) studied the influx of Na in single seminal

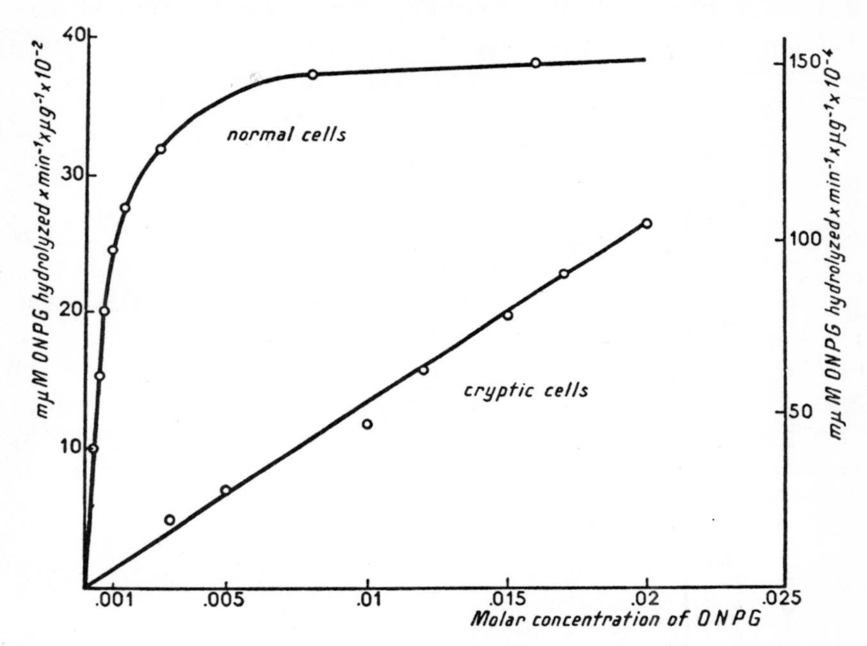

FIGURE 4–1 The influx of ortho-nitro-phenyl-β-D-galactoside (ONPG) in two types of *Escherichia coli* as measured by the rate of ONPG hydrolysis. Uper curve: normal type (left ordinate). Lower curve: cryptic (without permease) mutant (right ordinate). See als chapter 8 (Taken with kind permission from G.N.Cohen and J.Monod, 1957).

tubules. It can be seen that beyond 50 mM one of the two mechanisms saturates but the other keeps transfering more Na as the concentration of Na increases. Since the influx beyond 50 mM becomes a linear function of the concentration of Na, the increase seems to be due to simple diffusion. This situation is more commonly found than a penetration by simple diffusion only. In these circumstances simple diffusion is called the *leak.*

Since the driving force is proportional to the gradient of concentration, once C_2 becomes as high as C_1 the net flux vanishes. Therefore the maximal concentration that a non-charged substance could achieve inside the cell by simple diffusion is equal to the concentration in the bathing solution.

The critical factor in simple diffusion is the ability of the substance to diffuse in the membrane phase, so one would expect that, among substances which use this mechanism, the permeability will be proportional to its ability to penetrate lipids. We have already seen an example of this relationship when we discussed membrane structure (figure 1–1).

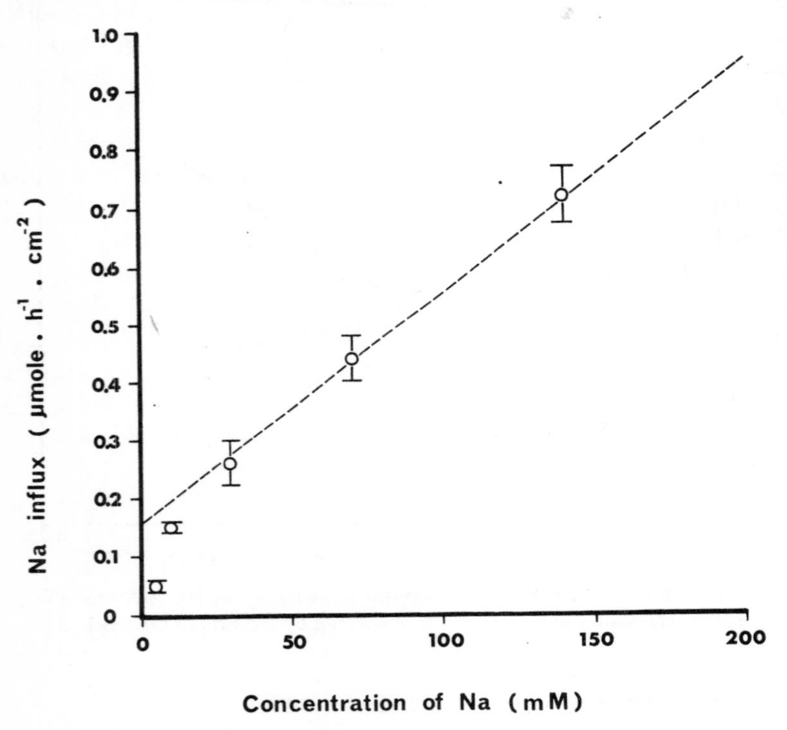

FIGURE 4–2 Influx of Na^+ as a function of sodium concentration in the bathing solution in single isolated seminal tube of the rat (Taken from I. Reisin and M. Cereijido, 1969).

4.2 *Facilitated diffusion*

Simple diffusion cannot explain all the peculiarities of the movement and distribution of substances that we mentioned in the introduction of chapter 3. Besides, if we plot in figure 1–1 the corresponding values of many of the most important metabolites, we will realize that they constitute remarkable exceptions to the rule that permeability depends on solubility in lipids. They enter into the cell at a rate far higher than their solubility in lipids would suggest. Glucose, for example, has five hydroxyl groups which form hydrogen bonds and would therefore require an activation energy in the neighbourhood of 20.000 calories per mole to diffuse across the lipid layer. The uptake of glucose would thus become a rare event. We know, though, that it is the other way around: glucose as most other metabolites, crosses the membrane quite easily. On general grounds if processes of simple diffusion only were involved, the membrane might be expected to be impermeable to sugar molecules with their hydroxyl groups and similarly to polyhydric alcohols, polycarboxylic and phosphoric acids, and to electrolytes. It was the necessity to explain the fast rate of permeation of biologically important substances, as well as the observed saturation of their fluxes at high concentration, and the incredible selectivity for certain substances that the concept of *facilitated diffusion* was developed (Davson and Reiner, 1942; Stein and Danielli, 1956).

The term "facilitated" arises from the comparison of actual permeabilities with what would happen if the substance had to diffuse through a lipid layer 50–70 Å thick. The real activation for glucose penetration into human or rabbit red cells for instance is around 4.000 calories, *i.e.* one fifth of the energy needed to break the hydrogen bonds by thermal agitation (Danielli, 1954). Facilitated diffusion is usually explained by assuming that the membrane has special entities (pores, carriers, dimerizers, etc.) and that the substance may use them to avoid the restriction to penetration imposed by the lipid component of the membrane. So any entity present in the membrane, which combines selectively with a given substance and forms a complex whose diffusion coefficient is higher than the diffusion coefficient of the free substance may facilitate the permeation. Proteins, polysaccharides and other substances which possess hydrogen-bonding groups could form a complex with the lipophobic solute and carry it through the membrane.

Facilitated diffusion is a passive mechanism in the sense that the driving force is provided by the gradient of electrochemical potential of the diffusing species under consideration or else, by the gradient of some other substance present to whose movement it is coupled (see countertransport, pag. 102).

As it always happens in Science, once somebody develops a concept, and it becomes crucial in explaining experimental observations, the idea can be traced back to the nineteenth century, then one discovers that the Greeks have already suggested it and, finally, the idea ends up as a quotation of the Old Testament. Since the concept of the cell is not that old, the germ of the idea of facilitated diffusion can so far be traced to Pfeffer (1890) only. The fact that this concept is almost as old as the concept of membrane itself reflects the concern of biologists in explaining just how can the cell tell a glucose or a given aminoacid from hundreds of other substances and make it enter into the cell a thousand times faster than another substance which might be chemically related but which is useless for the cell.

To the best of our knowledge, the hypothesis that a membrane could have carriers was put forward by Osterhout in 1930. The introduction of a carrier in the picture of the membrane brings along some other consequences, besides those of facilitating the diffusion. Permeation will now depend on how many carriers the membrane has, how strong is their affinity for the substance, how many molecules could use the same carrier at a time, whether there are other substances present which also possess affinity for the carriers, etc.

It is worth noticing that a pore, for instance, may also facilitate diffusion. Going back to figure 1–1 we notice that most exceptions (*i.e.* higher permeability than the oil-water partition coefficient would predict) are small molecules. This effect could be explained by assuming that the membrane has small holes (pores): molecules which are small enough to pass through the pores will avoid the hydrophobic component of the membrane. This point will be discussed in more detail after we learn something about the kinetics of facilitated diffusion, however it is important to keep in mind that a carrier is just one of the mechanisms that the membrane may have to speed up permeation.

4.3 *Kinetics of facilitated diffusion*

Even when carriers are not the only mechanism for facilitating diffusion, they offer more possibilities to explain membrane phenomena than any other mechanism proposed so far. It is therefore convenient to base our discussion of kinetics on the carrier model. Figure 4–3 illustrates one of the simplest models of carriers. It was proposed by Widdas (1952) to describe the transfer of glucose in the placenta of the sheep. The carrier is represented by the bold letter $\mathbf{C}$. The concentration of the substance to be carried is represented by C. The compartments at the two sides of the membrane are assumed to be reservoirs, so that C_1 and C_2 are constant. To avoid complications of the type illustrated in figure 4–2, *i.e.* when a substance penetrates by more than one mechanism we will assume that C only crosses the membrane when carried by $\mathbf{C}$. If the concentration of C on side 1 is higher than on side 2 ($C_1 > C_2$), and since the affinity is the same on both sides, the overall operation of this mechanism is this: molecules of C on side 1 reach the membrane and combine with $\mathbf{C}$ to form $C\mathbf{C}$ with a rate constant k_1

$$C + \mathbf{C} \underset{k_{-1}}{\overset{k_1}{\rightleftharpoons}} C\mathbf{C} \quad K_m = \frac{k_{-1}}{k_1}. \tag{4-1}$$

A single complex may reach side 2 by simple thermal fluctuation. As more complexes $C\mathbf{C}$ are formed, their concentration on side 1 of the membrane will make them migrate towards side 2. Once there, the concentration of C

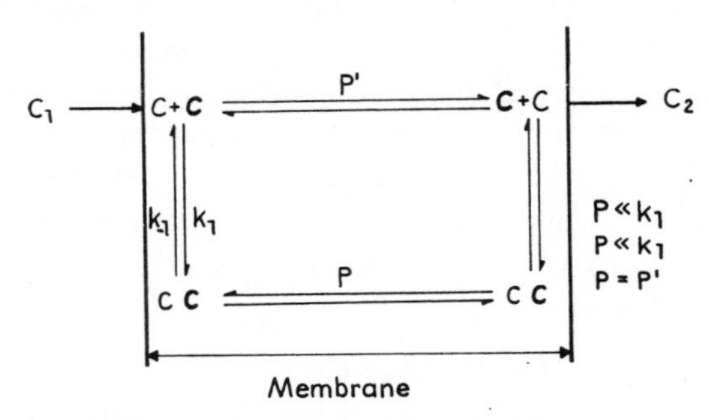

FIGURE 4–3 Schema of the carrier model proposed by W.F.Widdas, 1952.

is no longer high so according to reaction 4–1 the carrier will release the molecule of C. The concentration of free $\mathbf{C}$ on side 2, or thermal fluctuations will move the empty carriers toward side 1 and the whole cycle will recommence.

To simplify the explanation we will now make further assumptions. A convenient one is that the carrier moves through the membrane with the same ease when it is combined ($C\mathbf{C}$) as when it is free ($\mathbf{C}$). This is formally expressed by saying that the permeability of the free carrier (P') is equal to the permeability of the complex (P). We will also assume that the carrier is monovalent *i.e.* it will only translocate one molecule of C at a time. Another assumption will be that the rate constants k_1 and k_{-1} are relatively high and the permeability is relatively low so that the whole process is only rate limited by the trip across the membrane. At any given concentration there is a fraction (saturation fraction: θ_1) of the total carrier on side 1 ($\mathbf{C}_{t1}$) which is combined as $C\mathbf{C}$, and another fraction ($1 - \theta_1$) which is free as $\mathbf{C}$. A similar situation will exist on side 2 where the fraction of $\boldsymbol{C}$ combined as CC will be called θ_2.

Now we want to know how the flux J (a net flux) will depend on all the conditions stated above. Since J is a difference between J_{12} and J_{21} we have first to find out equations for these unidirectional fluxes. J_{12} will depend on the amount of complex on side 1 (expressed as combined carrier $\theta_1 \boldsymbol{C}_{t1}$) and the permeability P as follows

$$J_{12} = P\theta_1 \boldsymbol{C}_{t1}. \tag{4–2}$$

The amount of carrier combined will, in turn, depend on the velocity of formation, the velocity of dissociation, the velocity of escape of the complex to the other side of the membrane and the return from side 2 to side 1. Since we have already assumed that migration across the membrane is very slow as compared with the association and dissociation velocities, the escape and return of CC will not affect the amount of CC. This condition will permit the system to achieve equilibrium. The amount of CC will depend only on the velocities of formation (v_1) and dissociation (v_{-1}). The velocity of formation is a function of the availability of substance (C_1), free carrier $(1 - \theta_1)\, \boldsymbol{C}_{t1}$ and the rate constant k_1. Then

$$v_1 = k_1 (1 - \theta_1)\, \boldsymbol{C}_{t1} \cdot C_1. \tag{4–3}$$

The velocity of dissociation is a function of the amount of complex (expressed as combined carrier $\theta_1 \boldsymbol{C}_{t1}$) and the rate of dissociation k_{-1}

$$v_{-1} = k_{-1} \theta_1 \boldsymbol{C}_{t1}. \tag{4–4}$$

In equilibrium

$$v_1 = v_{-1}.$$

Solving equations 4–3 and 4–4 for θ_1 and since $Km = k_{-1}/k_1$, we obtain

$$\theta_1 = \frac{C_1}{C_1 + K_m}.$$

Introducing this value of θ_1 into equation 4–2

$$J_{12} = P\boldsymbol{C}_{t1} \frac{C_1}{C_1 + K_m}. \tag{4–5}$$

Using a similar reasoning we can derive an equation for J_{21}, then

$$J_{21} = P\boldsymbol{C}_{t2} \frac{C_2}{C_2 + K_m}. \tag{4–6}$$

The carriers, either free or complexed, migrate and distribute at random in the membrane phase; besides as it was assumed: $P = P'$. So we may take $C_{t1} = C_{t2} = C_t$ and use C_t in equations 4–5 and 4–6 instead of C_{t1} and C_{t2}.

Having derived equations for the unidirectional fluxes J_{12} and J_{21}, we may now combine equations 4–5 and 4–6 to obtain an expression for the net flux J

$$J = PC_t\left(\frac{C_1}{C_1 + K_m} - \frac{C_2}{C_2 + K_m}\right). \tag{4–7}$$

This equation, first derived by Widdas (1952) has since been proved to be a valuable tool in the study of the fluxes of many other substances across a wide variety of membranes. We will now illustrate the use of this formula and analyze some consequences of this treatment.

As mentioned above, facilitated diffusion saturates. This means that as we raise C_1, the flux will increase to a maximum value ($J_{\max}$). $J_{\max}$ will be achieved when the term between parenthesis of equation 4–7 reaches a maximum value. This maximum depends on a combination of a maximum value of $C_1/(C_1 + K_m)$ and minimum value of $C_2/(C_2 + K_m)$. As we increase C_1, the term $C_1/(C_1 + K_m)$ will approach 1. If we make C_2 equal to zero, $C_2/(C_2 + K_m)$ will be zero. Therefore the limiting value of the parenthesis term is 1 and is achieved when C_1 is high and C_2 is zero. Under this condition the net flux will be maximal and equation 4–7 will be

$$J_{\max} = PC_t. \tag{4–8}$$

So we observe that the maximum flux depends on the number of carriers present in the membrane and the ability of the complex to diffuse in the membrane phase.

We can replace the value of PC_t in equation 4–7 to obtain

$$J = J_{\max}\left(\frac{C_1}{C_1 + K_m} - \frac{C_2}{C_2 + K_m}\right). \tag{4–9}$$

Therefore, the carrier hypothesis visualizes saturation as a situation in which the concentration of substance is so high that all the carriers are busy translocating molecules across the membrane.

K_m is a measure of the apparent affinity of the carrier for the substance to be transported: as equation 4–1 shows, the higher the affinity of the carrier, the lower the K_m. A low value of K_m indicates that a relatively low concentration of substances will suffice to saturate the carrier system and obtain $J_{\max}$.

Let us now see how a high affinity between carrier and substrate (low K_m) and a simultaneous high substrate concentration affect the saturation fraction and the kinetics. The two conditions tend to saturate the carrier. The amount of free carrier will tend to be negligible and θ will tend to its maximum value, *i.e.* 1. This can be shown analitically through the following equation:

$$\theta = \frac{C}{C + K_m} = \frac{CC}{CC + \mathbf{C}}. \qquad 4\text{–}10a$$

As C becomes much larger than K_m, θ tends to 1. In the equation on the right we see that as θ tends to 1, $\mathbf{C}$ (the free carrier) becomes negligible. In this case we can use an approximate value of θ.

$$\theta \simeq 1 - \frac{K_m}{C} = \frac{CC - \mathbf{C}}{CC}. \qquad 4\text{–}10b$$

By using this approximation of θ instead of the definition made before, we can deduce equation 4–11 which describes the kinetic of the flux under *high saturation*

$$J \simeq PC_t K_m \left(\frac{1}{C_2} - \frac{1}{C_1}\right) \qquad 4\text{–}11$$

of the carrier. We notice that in this case, the flux is *directly* proportional to the K_m.

Figure 4–4 shows the relationship between the concentration (expressed as multiples of K_m) and the saturation fraction calculated as $C/(C + K_m)$ (solid line) or calculated with the approximation $(1 - K_m/C)$ (broken line). Notice that when the concentration reaches a value high enough as multiple of K_m the agreement between the two lines is very good and as consequence the net flux calculated with the saturation fraction given by the approximation also differs very little from the real one. When the gradient of concentration across the membrane is equal to $(C_1 - C_2)$ the net flux (figure 4–4) will be 14% of the J_{max} if calculated on the basis of the approximation while the real one is 10% of the J_{max}. At higher concentrations the difference is still narrower.

Let us discuss the opposite case, *i.e.* when K_m is very high $K_m \gg C_1$; $K_m \gg C_2$. In this case we can transform equation 4–9 and neglect all terms in the common denominator which contain C_1 and C_2 to obtain:

$$J \simeq \frac{J_{max}}{K_m}(C_1 - C_2). \qquad 4\text{–}12$$

We notice that the flux is inversely proportional to K_m and the kinetics look like the one of a mechanism of simple diffusion. The low affinity of the carrier makes it necessary to use a very high concentration of substance to reach

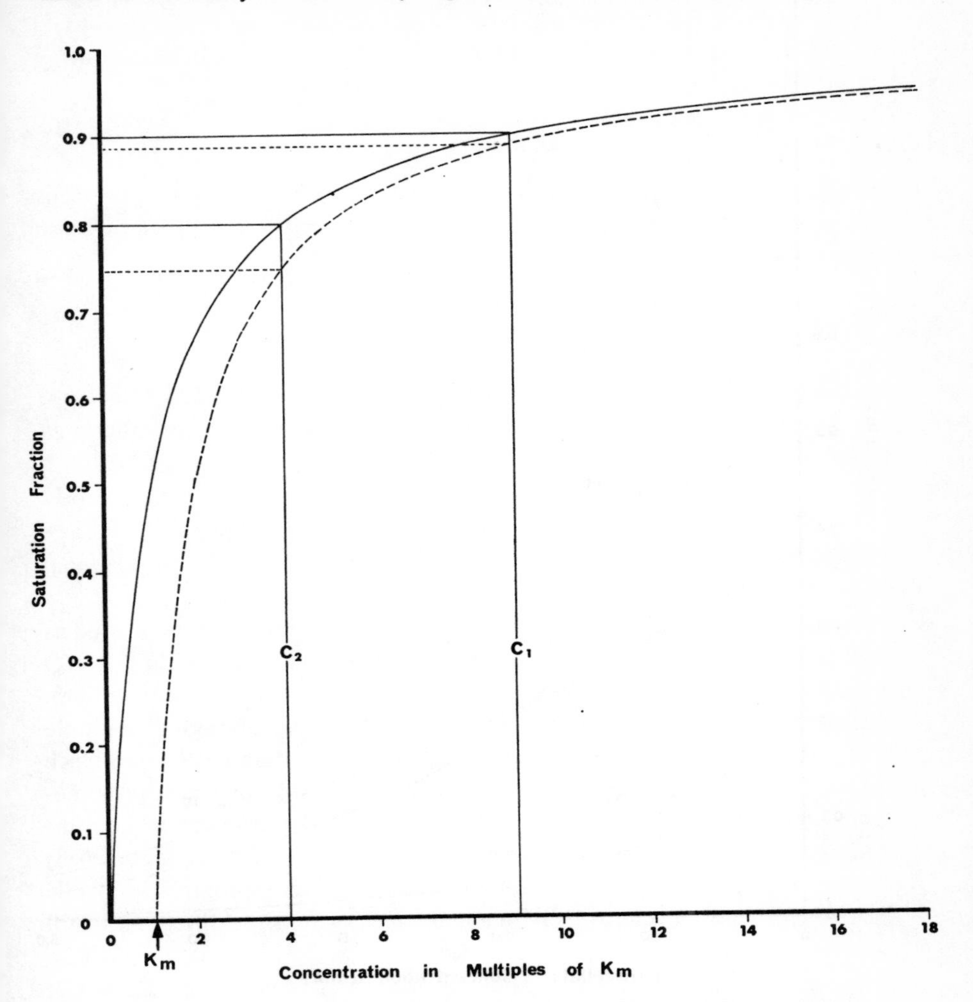

FIGURE 4–4 Effect of concentration on saturation fraction $(C/C + K_m)$ (solid line) and the approximation $(1 - K_m/C)$ (dotted line) from which equation 4–11 is derived. Note how the approximation improves at high concentrations. Net transfer depends on the difference in saturation due to C_1 and C_2 (Courtesy of W.F.Widdas).

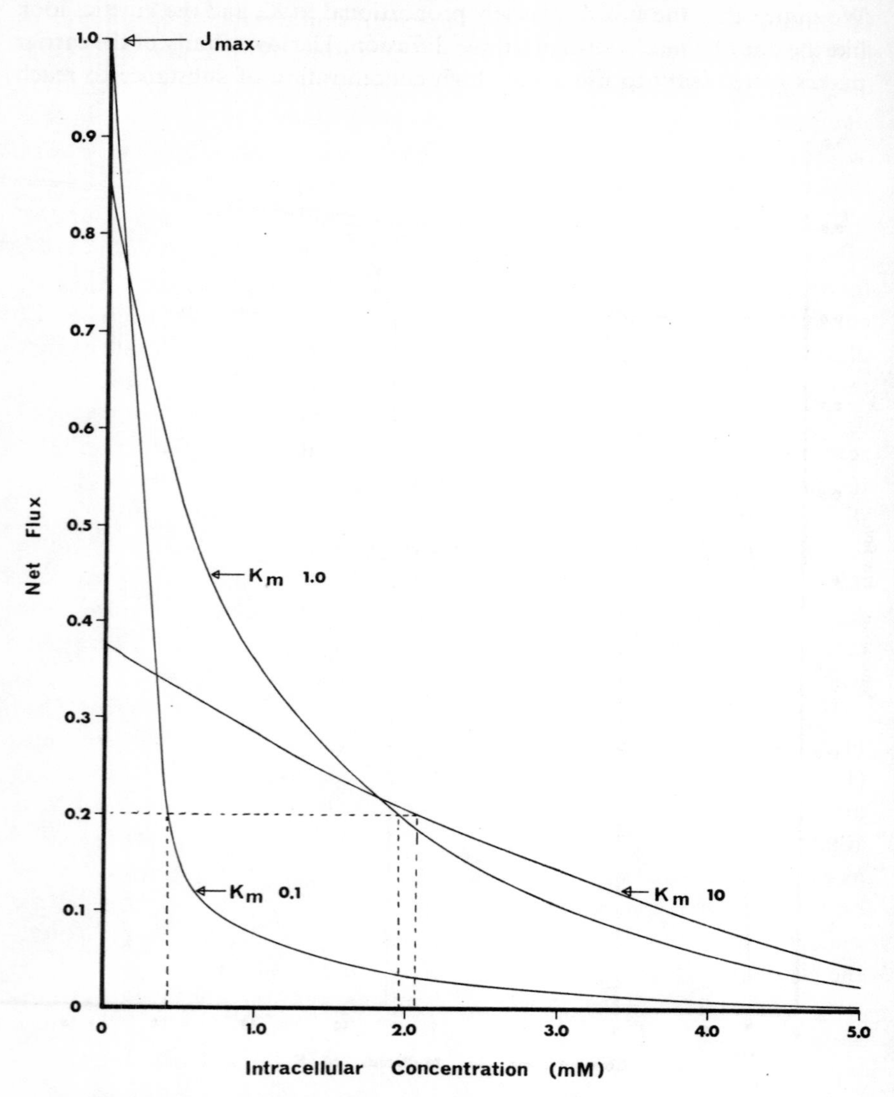

FIGURE 4–5 Net flux of glucose from plasma (taken as 6 mM) into cells containing glucose (0–5 mM) showing effect of K_m (0,1, 1,0, and 10 mM). Dotted lines indicate steady state cell concentration if metabolism removes glucose at the rate of 0.2 J_{max}. (Courtesy of W. F. Widdas).

saturation. Sometimes the necessary concentration to reach this point is so high that, if used, it would damage the preparation. With ketose sugars for instance it is virtually impossible to do experiments at concentration which show saturation. (Le Fevre and Davies, 1951; Widdas, 1954.) In this case detection of the carrier system, instead of being based on saturation, has to be based on some other properties of the carriers.

To illustrate further the effects which the affinity of the carrier for the substrate may have, consider figure 4–5 which illustrates the net flux of, for example, glucose into cells when provided by a source of nearly constant concentration such as the plasma (here taken as 6 mM for purposes of illustration). It will be noted that if the K_m value is low (high affinity) the net flux, although starting at a high level, quickly reduces to a very small value as the internal concentration rises. This is because efflux quickly approaches saturation and balances the influx. With higher values of K_m (lower affinity) the net flux starts at a lower fraction of J_{max} but falls off more slowly as the internal concentration is raised.

Note that if metabolism could remove glucose at a rate about equal to $0.2\,J_{max}$, the intracellular concentration could never exceed 0.5 mM when K_m is equal to 0.1 but could reach about 2 mM in either of the other two cases. This illustrates the importance of considering both influx and outflux when studying permeation.

Figure 4–6 shows how net flux depends on K_m. Substance A ($K_m = 5$) has higher affinity for the carriers than substance B ($K_m = 100$). The net influx (1 to 2) is explored in a wide range of concentrations. The concentrations are arranged in such a way that C_1 is always 10 times higher than C_2; J_{max} is 100. Notice: a) In each curve, when C_1 and C_2 are much lower than K_m, as the concentration rises the net influx goes up, b) when the concentration on side one of the membrane is high enough, the influx saturates but the outflux keeps rising, so that the difference (net flux) decreases (right limb of the curves). c) at low concentration (say $C_1 = 10$) substance A, which is the substance with the higher affinity ($K_m = 5$) has also the higher net flux (low saturation case), d) at higher concentration (say $C_1 = 10^2$) the substance with the higher affinity has the lower flux (high saturation case), e) since the concentration on side 2 is never zero, there is always an outflux, therefore the net influx never reaches J_{max}.

In some experiments it is possible to measure the individual fluxes. In particular, when the substance can be labelled with a radioisotope, to study the system at the initial time (when C_2 is zero). In this period the outflux

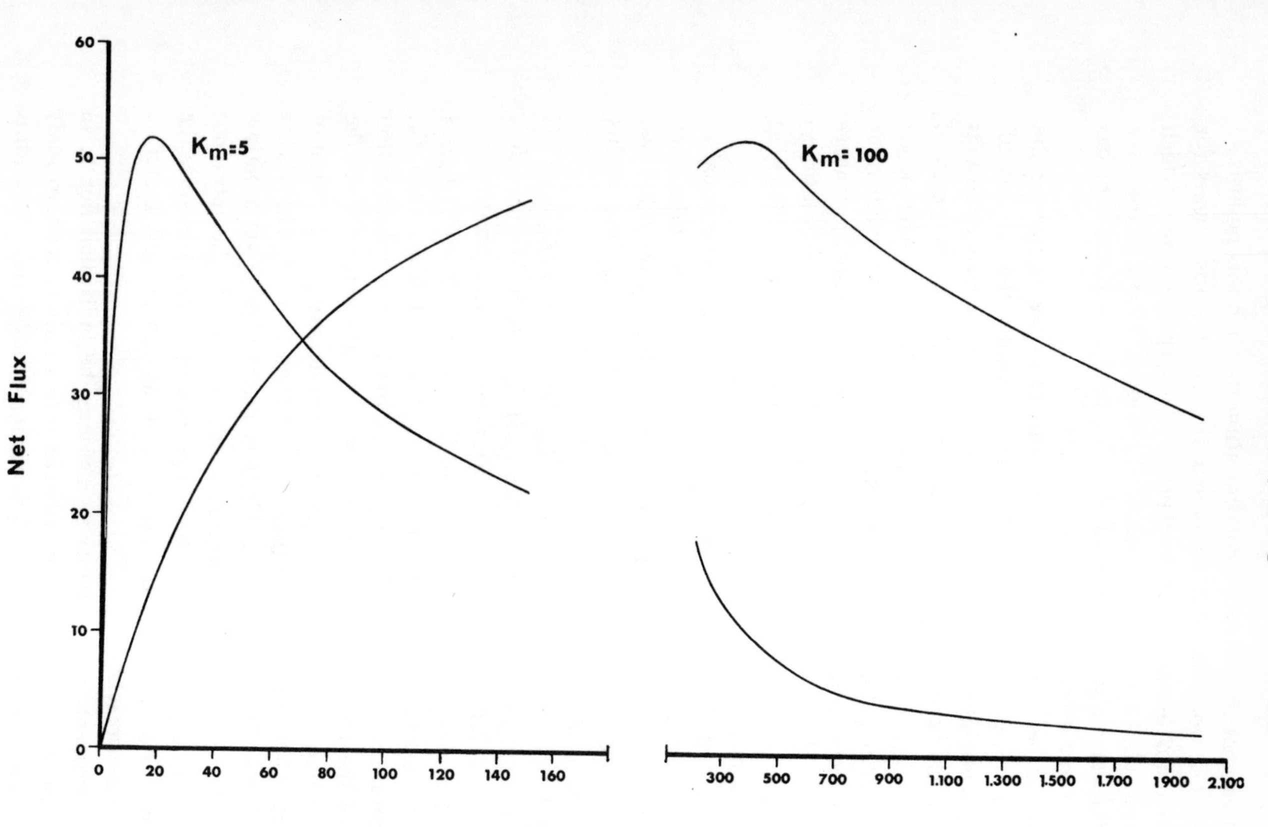

FIGURE 4–6 Net influx of two different substances, each one studied separately as a function of concentration. The experimental conditions are arranged so as to keep C_1 (concentration in the bathing solution) ten times higher than C_2 (concentration in the cells). Notice the change of the scale of the abscissa after 160 mM.

(J_{21}) is negligible and equation 4–9 may be converted into

$$J = J_{12} = \frac{J_{max} C_1}{C_1 + K_m}. \qquad 4\text{–}13$$

This formula can be arranged as follows

$$\frac{1}{J} = \frac{K_m}{J_{max}} \frac{1}{C_1} + \frac{1}{J_{max}}. \qquad 4\text{–}14$$

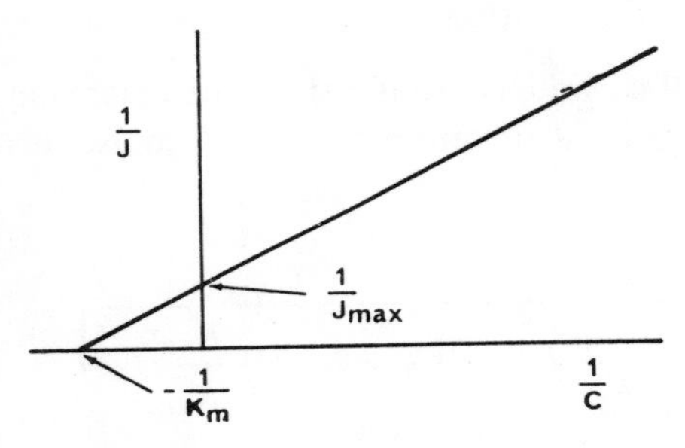

FIGURE 4–7

Therefore, if we measure the initial influx at several concentrations and we plot in a Lineweaver-Burk's graph (fig. 4–7) we can obtain the values of J_{max} and K_m. If instead of studying uptakes we run a wash-out experiment (see chapter 2) we can measure J_{21} using the same approach.

4.4 Inhibition

Since facilitated diffusion by a carrier mediated mechanism depends on the formation of the complex carrier-substance, an impairment of the combination of the substance with the carrier will affect the flux. This impairment can be produced, basically, by two different groups of inhibitors: competitives and non-competitives.

4.4.1 Competitive inhibition

A substance (C) complexes with a carrier **C** when its stereochemical properties fulfill the requirements of the carrier. It may be the case that another substance (I) would also meet these requirements. If they are present at the same time, they will compete with each other for the use of the carrier and their fluxes will be lower than they would be if measured in separated experiments. Substance I will be said to produce a competitive inhibition of the flux of C (we could say just the same of C with respect to the flux of I). Substance I will combine with $\mathbf{C}$ according to

$$I + \mathbf{C} \rightleftharpoons I\mathbf{C}.$$

So that the carrier $\mathbf{C}_t$ may be found in three different conditions: free $\mathbf{C}_f$, combined with C and combined with I

$$\mathbf{C}_t = \mathbf{C}_f + C\mathbf{C} + I\mathbf{C}.$$

To derive the equation for the flux, we have to use the same reasoning that led us to equation 4–7. However, in the present case we have to use this new value of $\mathbf{C}_t$ to obtain

$$J = P\mathbf{C}_t\left(\frac{C_1}{C_1 + K_{mc} + \dfrac{K_{mc}}{K_{mi}}I_1} - \frac{C_2}{C_2 + K_{mc} + \dfrac{K_{mc}}{K_{mi}}I_2}\right) \qquad 4\text{–}15$$

where K_{mc} and K_{mi} referred to the K_m of substances C and I respectively. K_{mc} of equation 4–15 is the same as the K_m that can be determined in an experiment where the inhibitor is absent. If for a particular system K_{mc} does not coincide with the K_m calculated in the abscence of inhibitor, it would indicate that the validity of the assumptions made to derive these formulas do not hold for the system. For real cases in which this check was made, see Sen and Widdas (1962) and Rickenberg and Maio (1960).

The general practice is to study the flux of C as a function of its concentration both, with and without I, which is used at a constant concentration. When studied under this condition, the presence of the inhibitor seems to raise the K_m. According to equation 4–8 $J_{\max}$ depends on the density of the population carriers and the permeability of the carrier. None of these parameters is affected by the inhibitor. Therefore the value of $J_{\max}$ remains unchanged in the presence of the inhibitor. Of course the necessary concentration of substance C to reach a maximum flux will be higher than in the abscence of the inhibitor because C has to compete with I for the use of the carrier. Since the total number of carriers (as expressed by $J_{\max}$) and the permeability are not changed, the fact that it takes a higher concentration of C to saturate them gives the impression that the carriers have lost in part their affinity for the substance. The presence of an inhibitor rises the value of the apparent K_m. Note that this K_m is actually an *apparent* K_m, *i.e.* an operational parameter. The *real* K_m (*i.e.* k_{-1}/k_1) is a thermodynamic constant. This is one of the cases in which the thermodynamic and the apparent value of the K_m do not coincide. Figure 4–8 shows an example of inhibition. (Nathans, Tapley and Ross, 1960.) It illustrates a study of the flux of L-monoiodotyrosine (MIT) through a membrane of rat intestine. The flux of MIT was studied as a function of concentration under two different

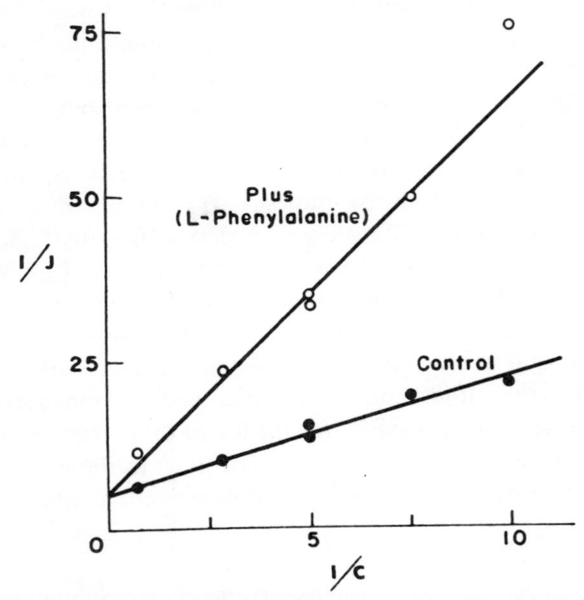

FIGURE 4–8 Competitive inhibition of the flux of L-monoiodotyrosine by L-phenylalanine in the intestine of the rat. Notice that both curves intercept the ordinate at the same point (Taken with kind permission from D. Nathans, D. F. Tapley and Ross, J.E., 1960).

conditions: when it was alone (control) and when there was an inhibitor (L-phenylalanine) present. Compare this figure with figure 4–7 and notice that the presence of the inhibitor does not alter the value of J_{max}, but it shifts the K_m toward a higher value (lower affinity).

As we see, the key of the competitive inhibition is the occupancy of the carrier by a molecule of inhibitor, because it diminishes the interaction between the substance C and the carrier. It is not essential for the inhibition that the complex carrier-inhibitor could diffuse through the membrane.

It follows that a substance may have affinity for the carrier, form a non-permeant complex and still act as an inhibitor. This competitive inhibition by substances that are not permeant gives rise to very curious phenomena. Let us analyze, for instance, the following example. Phloridzin and its aglycone Phloretin produce a competitive inhibition of the influx of glucose into human red blood cells (Wilbrandt, 1950). Many molecular species having as a common feature a phloretin group can inhibit the flux of glucose (Le Fevre, 1961). One of them, polyphloretin phosphate (PPP) which is itself a non-permeant substance, has the peculiar property of inhibiting more the *outflux* than the influx (Wilbrandt, 1954). It is somewhat puzzling that an outfluxing glucose, that has to "board" the carrier at the membrane-cytoplasm boundary, should be inhibited by a substance which can only interact with the carrier on the other side of the membrane. A plausible explanation of this effect is the following: when one studies the *influx* both, glucose and PPP are added to the outside and, if glucose is used at sufficiently high concentration, it can displace PPP from the carriers and achieve a J_{max}. When the cell is preloaded with glucose to study its *outflux*, and PPP is added to the bathing medium, it also forms a PPP-C complex which, because of its non-permeability, accumulates at the outer solution-membrane boundary. This time though, glucose in the bathing solution is too diluted, and glucose in the cell is on the wrong side to challenge PPP for the use of the carrier. Thus the outflux is severely reduced because the carriers are stuck at the wrong side and there is in fact no real "competition" (see Bowyer and Widdas, 1956).

4.4.2 *Non-competitive inhibition*

When the combination carrier-inhibitor is too strong, it cannot be modified by rising the concentration of substance *C*. It could also occur that the combination carrier-inhibitor takes place at a site of the carrier which is different from the site where the carrier combines with *C* and competition is altogether impossible. Several *SH*-blocking agents inhibit the transport of sugars in the red cells in a non-competitive way (Le Fevre, 1948; Wilbrandt and Rosenberg, 1950; Dawson and Widdas, 1963). This fact, by the way, suggests that somehow proteins participate in the movement of glucose. The uranyl ion (Rothstein, 1954) p-chloremercuribenzoate (Battaglia and Randle, 1960) are just a few example of substances that can produce non-competitive inhibition.

Since the combination carrier-inhibitor is not competed against, each time one of these complexes is formed, a carrier is functionally cancelled. This reduction of the total number of carriers available, confers on this type of inhibition its most characteristic feature; the reduction of the value of J_{max} (see equation 4–8).

4.5 *Specificity of the facilitated diffusion*

Biological membranes posses an exquisite specificity for certain substances. The origin of this selectivity was looked for in the size, shape, lipid solubility, chemical structure, type of isomorization, charge, etc. of the selected molecule. We will refer now to the approaches used to study the selectivity of the carriers system and give some idea on the basis of this discrimination.

Since J_{max} depends on the population of carriers, one suspects that a group of substances that show the same J_{max} use the same carrier system, because it would be unlikely to have two completely different systems with the same number of carriers. If they are inhibited by the same inhibitors we have more grounds for suspicion. If we further prove that the substances belonging to the group compete with each other when put together, we can safely assume that they are using the same carrier system. Now we are in the position to study what is the cause of selectivity.

The principal indicator is the K_m because it is a function of the affinity between the carrier and the substance. So one can make a ranking of selectivity on the basis of the K_m and then try to associate this order with some feature of the permeant molecule. Thus Le Fevre and Marshall (1958) (see also Le Fevre and Davies, 1951 and Le Fevre, 1961) studied the facilitated passage of monosaccharides through the membrane of human erythrocytes and observed that the decreasing order of selectivity is: 2-deoxy-D-glucose, D-glucose, D-mannose, D-galactose, D-xylose, L-arabinose, D-ribose, D-lyxose, D-arabinose, L-fucose, L-rhamnose, L-xylose, L-galactose, L-glucose. They pointed out that this order suggests that the affinity is higher as the pyranose ring of the sugars tends to assume the particular "chair" shape designated by Reeves (1951) as "C_1 conformation". Among the group of sugars that meet these requirements, the affinity for the carriers is higher as the number of –OH or $-CH_2OH$ groups orientated in the equatorial position is larger. An idea of the difference of affinity for the different sugars is given by the fact that, in order to achieve one half of the J_{max}, it takes a concentration of L-glucose higher than 3 M, while it takes only 0.005 M 2-deoxy-D-glucose to do the same. The sugar selectivity was also studied in foetal erythrocytes by Widdas (1955).

The system which facilitates the diffusion of sugars in the intestine is less specific. It requires a pyranose ring and a non-substituted –OH in the position which corresponds to carbon 2 in D-glucose (Crane, 1960). Accordingly, this system does not translocate fructose, 2-deoxy-D-glucose nor 2-deoxy-D-galactose, which are instead readily handled by the carriers in the erythrocytes.

Selectivity is often studied by comparing the ability of different molecules to inhibit competitively not each other fluxes, but the flux of a third substance. Thus Fukui and Hochster (1964) studied the competition exerted by D-galactose, D-fucose, D-xylose and D-glucose on the uptake of C^{14}-sucrose by cells of *agrobacterium Tumefaciens*. Figure 4–9 shows the effect of glucose

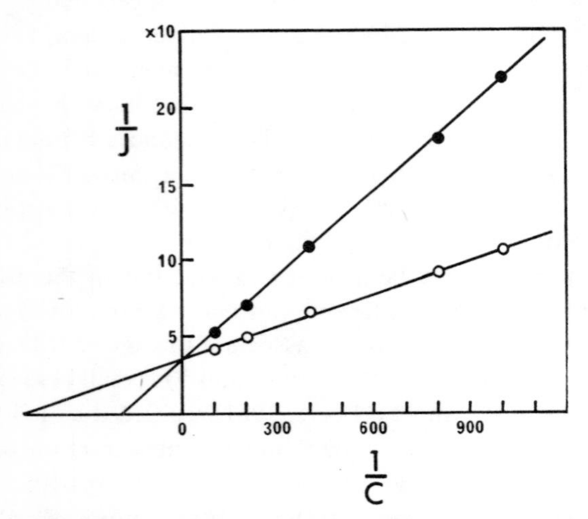

FIGURE 4–9 Influx of sucrose: control (open circles), and in the presence of glucose (full circles) (Taken with kind permission from, S. Fukui and R. M. Hochster, 1964).

3×10^{-2} M on the sucrose flux. It can be observed that glucose shifts the value of the apparent K_m of the sucrose without modifying the J_{max}. Figure 4–10 shows the effect of several concentrations of inhibitor on the uptake of sucrose. Notice that the uptake of sucrose was studied at two different concentrations of sucrose (2.5 and 5.0 mM). It can be shown that the value of the abscissa at the point where the two curves cross each other gives the value of the dissociation constant of the inhibitor. This can, in turn, be used to compare the selectivity for different inhibitors.

For the sake of clarity we have restricted our description of facilitated diffusion to the model of carriers and, among carriers, to a very simple version. Now that we have an idea of how a carrier works, we can figure

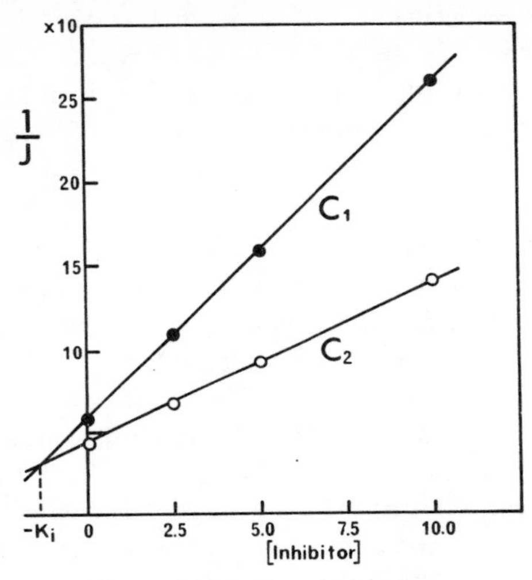

FIGURE 4–10 Influx of sucrose from two different concentrations of sucrose in the presence of an competitive inhibitor. Open circles 5.0 mM sucrose. Full circles 2.5 mM sucrose (Taken with kind permission from S. Fukui, and R. M. Hochster, 1964).

carriers which are not monovalent; or carriers which can—to a certain extent—leave the membrane; or kinetics in which the relationship between free carrier/complexed carrier at the membrane-solution boundaries do not achieve equilibration; or carriers that, when combined with the substance, travel faster than in the free state; or carriers with allosteric properties; or carriers that have to combine with two different substances (usually one of them being Na^+) to achieve their maximal efficiency. Besides, a given substance may use more than one system to cross the same membrane, a situation which obviously complicates the study. (Figure 4–2 is an example. See also Battaglia and Randle, 1960.) For example of carriers with different properties than the one described here see Rosenberg and Wilbrandt (1955 and 1963); Stein (1962b); Rickenberg and Maio (1960), Jacquez (1961, 1963 and 1964); Dawson and Widdas (1964); Wong (1965); Schultz, Curran, Chez and Fuisz (1967); Curran, Schultz, Chez and Fuisz (1967); Chez, Palmer, Schultz and Curran (1967).

4.6 Some consequences of the carrier model

Usually the phenomena we are going to describe in this section are not just discovered in the course of an experiment, but they are provoked by the investigator when he suspects that the mechanism involved is a carrier, because the best explanation we have for them, as yet, is based on carrier models. We will describe two of these phenomena: exchange diffusion and countertransport.

4.6.1 Exchange diffusion

Exchange diffusion is easily visualized by assuming a carrier system that —because of its hydrophylic groups—cannot traverse the membrane as a free carrier and can only diffuse when the carrier is complexed*.

Formally this means that P' in figure 4–3 is zero. The carrier will transfer a molecule from one side to the other and stay there until another molecule brings it back. This system does not produce a net flux even if it had a concentration difference between both sides of the membrane. The carrier only exchanges molecules between side 1 and side 2. This is easily detected by adding a labelled substance to side 1. When the carrier transports a labelled molecule from side 1 to side 2 and releases it, there is a very small chance that if will combine again with a labelled molecule. So the trip from side 2 to side 1 is likely to be made with a non-labelled molecule. In this way we will observe a large unidirectional flux in abscence of a net flux. Notice that in this particular case, if the concentration on side 2 were zero, not only the net flux but the tracer flux as well will be zero.

The basic idea of exchange diffusion is, therefore, the combination of large unidirectional fluxes with a small net flux. It is not necessary, though, that this condition arises from the low permeability of the free carrier (see also Rosenberg and Wilbrandt, 1957).

Le Fevre and McGinnis (1960) have compared equations to describe glucose permeation in red blood cells under two different conditions: a) penetration by simple diffusion (fig. 4–11 left) and carrier mediated facilitated diffusion (fig. 4–11 right). Notice that the sugar content rises at about the same speed in both cases. The unidirectional fluxes, though, are much faster in the facilitated diffusion case. To test this model Le Fevre and McGinnis

* in the case of ion permeation this is due to electroneutrality requirements (see chapter 6).

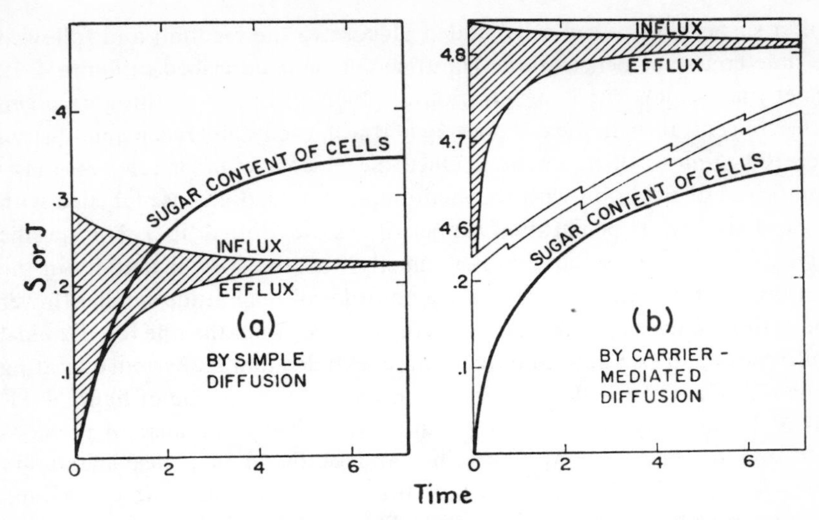

FIGURE 4–11 Alternative theoretical patterns for glucose fluxes in red blood cells. Ordinate: cell sugar content (S) or fluxes (J). Abscissa: time. The unit time used is the half-equilibration time (Taken with kind permission from P. G. LeFevre and G. F. McGinnis, 1960).

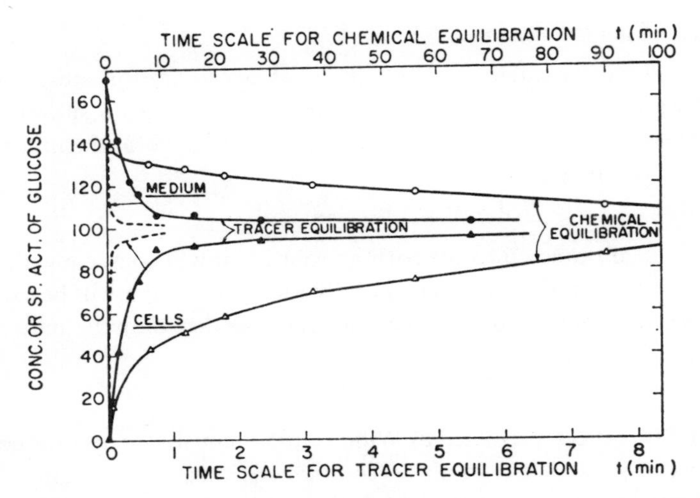

FIGURE 4–12 Time course of the glucose and C^{14}-glucose penetration into red cells. See text (Taken with kind permission from P. G. Le Fevre and G. F. McGinnis, 1960).

took a suspension of red cells, added glucose to the medium and followed the time course of penetration. The time course is described in figure 4–12 (upper time scale). The concentration of glucose in the medium goes down as the concentration in the cell goes up and will, eventually reach equilibrium after 100 minutes of incubation. Once the glucose content reaches quasi-equilibrium distribution with the medium, they added glucose labelled with C^{14} and studied its penetration (upper time scale, dotted line). The specific activity of C^{14}-glucose equilibrates quickly with the specific activity in the solution. Even plotting in a time scale an order of magnitude smaller (lower time scale) the tracer equilibration curve is steeper than the one for chemical equilibration. So Le Fevre and McGinnis concluded that the system operating in the red cell is of the type described on the right hand side of figure 4–11. This is, by the way, a case of high saturation kinetic: because of the concentration of glucose used, the influx and outflux had a near-maximum value. So, shortly after the start of the experiment there were maximal unidirectional fluxes and low net flux. This explains the fast equilibration of tracer (which depends on the exchange) and the slow accumulation of glucose (which depends on the net flux). This, also illustrates a case of exchange diffusion due to an exhaustion of free carriers, and not to the impermeability of the free carrier illustrated before.

4.6.2 *Counter-transport*

Imagine a cell with a carrier system which has selectivity for substances A and B. If we incubate the cell with a high concentration of substance A (B is not present) and the system reaches a point where both, the influx and the outflux saturate, then

$$J_{12}^{A} = J_{21}^{A}; \quad J_{\text{net}}^{A} = 0.$$

If we now add substance B to the bathing solution it will compete with substance A for the use of the carrier. Initially this competition will be exerted only at the outer solution-membrane boundary, so that only the influx will be affected. Therefore

$$J_{12}^{A} < J_{21}^{A}; \quad J_{\text{net}}^{A} \neq 0.$$

In summary, this effect, predicted by Widdas (1952), consists of the production of a net flux of substance A in abscence of a concentration gradient of this substance and as a result of the competition by substance B. It is called "counter-transport" because the net flux goes in opposite direction to the flux of the substance which produces it. Notice that, once substance B achieves equilibrium on both sides, J_{21} will be competed with as much as J_{12} and the net flux will vanish. Notice also that substance B might be metabolized in the cell and might never reach in the cell a concentration as high as in the bathing solution. The most general case is given when substance A is a radioactive tracer of substance B.

Park, Post, Kalhan, Wright, Johnson and Morgan (1956) observed countertransport of xylose in rabbit red cells when glucose was added to the medium. Burger, Hejmova and Kleinzeller (1959) observed countertransport of galactose and arabinose in yeast cells treated with glucose. For other examples read Rosenberg and Wilbrandt (1957) and Cirillo (1960).

4.7 *Models*

As explained above, we have chosen to describe facilitated diffusion on the bases of the carrier hypothesis because carrier models are very versatile. Carriers, though, are just one of the possible ways of explaining the facilitation of the diffusion of substances through a lipid film. Pores provide another way of facilitating permeation through this film.

4.8 *Pores*

Figure 1–1 shows that small molecules cross cell membranes faster than expected on the basis of their solubility in lipids. This could be accounted for by assuming that the membrane has cracks through which the molecules could pass provided they are small enough. This is illustrated in figure 4–13: as the size of the molecule is bigger its ability to cross the membrane decreases. In this case, though, the biological membrane involved is not a simple cell membrane but one of the type described in chapter 7. Another support for the pore theory comes from the possibility of eliciting electrokinetic effects. These effects, due to coupling between water and ion movement, will be studied in chapter 6.

4.8.1 *Existence and shape of the pores*

The classical idea of a pore is that of a cylindrical hole which remains static and which is perpendicular to the surface. Modern views differ considerably from this picture. Pores are thought to be a consequence of the thermal energy of the molecules in the membrane and in the solutions. In a given point of the membrane the molecules may move appart producing a gap big enough to let molecules of solvent and solute through. The membrane structure is also continuously bombarded by the molecules in the solutions. A small, but not negligible, fraction of these collisions may have a kinetic energy high enough to cause a local disruption. We also have to keep in mind that the bimolecular leaflet model of the membrane seems to be just one of

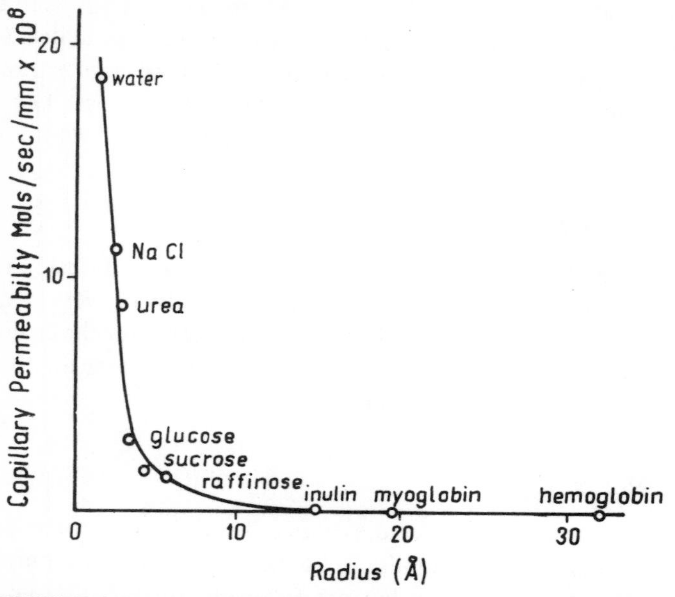

FIGURE 4–13 Permeability *vs* molecular size o fthe permeating molecule (Taken with kind permission from J.R.Pappenheimer, 1953).

the possible states in which the membrane can exist. If lipids can adopt in the membrane the same configurations they show in lipid-water systems (see for instance figure 1–27b), they could probably form spontaneous pores across the membrane. In fact Muller and Rudin have suggested that some substances (see page 145) can speed up permeation by "micellating" the membrane. Thermal fluctuations might create a pore *ad hoc* for the permeating molecule. For the benefit of non-specialists, Dr. Villegas from Caracas described this process by making a comparison that we are tempted to quote here: "Take a platter full of spaghetti. Put a marble on top. Shake the platter. After a while the marble (which is quite insoluble in spaghetti) will reach the bottom. No pore was observed before or after the experiment. This process can be compared to pore facilitated diffusion."

Static pores which remain continuously in the same place were also postulated. Figure 4–14 illustrates a pore proposed by Danielli (1954) which is lined by the protein envelope of the membrane. All these pores may have electric charges fixed to their walls. Pores with electric charges play a major role in explaining ion translocation and electrical phenomena, therefore the subject will be retaken in chapter 6.

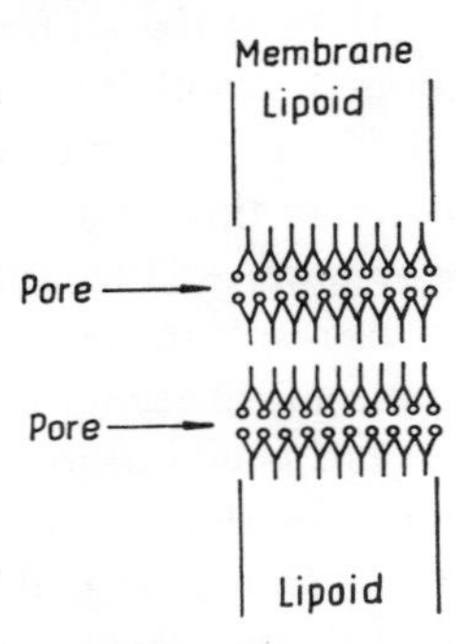

FIGURE 4–14 Pore in a lipid membrane (After J.F.Danielli, 1954).

In summary, pores are standing or transient holes of any shape, follow a straight or tortuous path across the membrane, and their walls may have charges or groups with affinity for certain substances.

4.8.2 The size and number of pores

Regardless of whether they are static or ephemeral entities, the sum of the area of all the gaps in the membrane of a cell will have an almost constant average value. That is called "the pore area". It is important to know whether this area is afforded by a few big pores or a multitude of narrow pores. Studies in this field try to stablish how many cylindrical holes, of a given radio and a length equal to the thickness of the membrane would be needed to account for the pore area. Early attempts were based on the measurements of the water flow under a pressure gradient and the diffusion of water under zero pressure gradient using labelled water. Assuming that the first process obeys Poiseuille's law and the second obeys Fick's law, the information obtained with the two sets of experiments permits to solve for the radius of the pore. This approach rests on the extrapolation of macroscopic laws down to the molecular level and therefore requires considerable corrections. For detailed discussion of the rational methods and criticism see Pappenheimer, Renkin and Borrero (1951); Koefoed-Johnsen and Ussing (1953);

Sidel and Solomon (1957); Paganelli and Solomon (1957). To avoid these problems another approach was devised. It is based on the use of Renkin's equation, Staverman's reflection coefficient and the Durbin-Frank-Solomon's equation. We will then discuss these briefly and then see how they are used to measure pores.

1) *Renkin's equation (1954)*. Assume a molecule barely smaller than the pore. In order to penetrate this molecule has to be aimed right at the central axis of the pore, otherwise it will hit the border and rebound. Another molecule, smaller than the first, could penetrate the pore even if it is not aimed at the axis. So there will be a small area around the axis where this second molecule can arrive and still miss the rim and get into the pore. As the molecule gets smaller and smaller the area where the hits will result in penetration gets larger and larger. When the radius (a) of the molecule is negligible with respect to the radius (r) of the pore, the effective area of penetration will be equal to the pore area (A_p) and this, of course, holds for the effective area for the molecules of solvent (A_w) as well as for the solute (A_s). Renkin derived formulae relating these areas to the radius of the molecules and the radius of the pore

$$\frac{A_s}{A_p} = \left[2\left(1 - \frac{a_s}{r}\right)^2 - \left(1 - \frac{a_s}{r}\right)^4\right] \times$$
$$\times \left[1 - 2.104\left(\frac{a_s}{r}\right) + 2.09\left(\frac{a_s}{r}\right)^3 - 0.95\left(\frac{a_s}{r}\right)^5\right]$$

where A_s is the effective filtration area for the solute, and a_s is the radius of the solute molecule. A similar equation can be written for A_w, the effective filtration area for the water and using a_w, the radius of the water molecule instead of a_s. The ratio between A_s/A_w is:

$$\frac{A_s}{A_w} = \tag{4–16}$$
$$\frac{\left[2\left(1 - \frac{a_s}{r}\right)^2 - \left(1 - \frac{a_s}{r}\right)^4\right]\left[1 - 2.104\left(\frac{a_s}{r}\right) + 2.09\left(\frac{a_s}{r}\right)^3 - 0.95\left(\frac{a_s}{r}\right)^5\right]}{\left[2\left(1 - \frac{a_w}{r}\right)^2 - \left(1 - \frac{a_w}{r}\right)^4\right]\left[1 - 2.104\left(\frac{a_w}{r}\right) + 2.09\left(\frac{a_w}{r}\right)^3 - 0.95\left(\frac{a_w}{r}\right)^5\right]}.$$

2) *Staverman's reflection coefficient* (σ) *(1951)*. Consider a cell that, although immersed in pure water, swells but does not burst (fig. 4–15). If we add a

non-permeable solute to the bathing solution it will elicit an osmotic outflux of water and the volume of the cell (V) will decrease (curve 4). The time course of this change will depend on the water permeability (descending limb of the curve) and the osmotic pressure excerted by the solute added. This osmotic pressure may be calculated by van't Hoff equation ($\pi_{\text{theor.}}$). If the molecules of the solutes added were somewhat smaller than the pores of the membrane, they will also shrink the cell, but the molecules of solute will

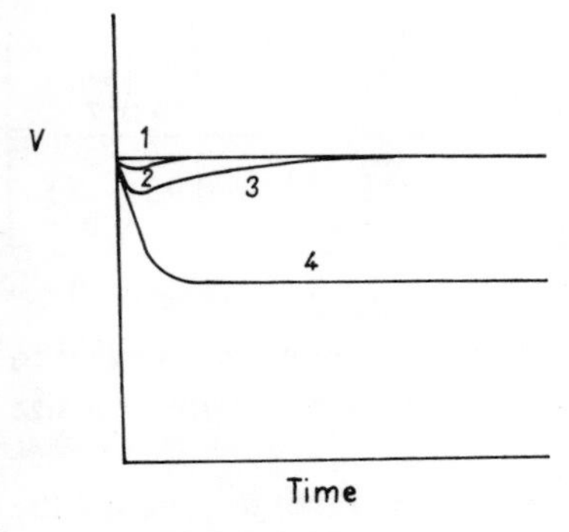

FIGURE 4–15

penetrate the cell before an equilibrium is reached (curve 3). Therefore, this solute will never exert the osmotic pressure as predicted by the van't Hoff equation but a lower one (π_{observ}). Further penetration of solute will cancel the osmotic gradient imposed and the cell will recover its initial volume. Curve 2 illustrates the case of a solute with even smaller molecule. The change of the volume tends to disappear as the size of the molecules added is closer to the size of the molecules of solvent. If we add more water, for instance, no change at all will be observed (curve 1). The relationship between π_{observed} and $\pi_{\text{theoretical}}$ is called the "reflection coefficient" (σ).

$$\sigma = \frac{\pi_{\text{obser}}}{\pi_{\text{theoret}}}.$$

In summary: the change in osmotic pressure exerted by a solute depends on the relative size of its molecule, the molecule of the solvent and the diameter of the pore. The osmotic pressure developed could be as high as the theoretical ($\sigma = 1$) when the solute is larger than the pore or nil ($\sigma = 0$) when the molecule of solute compares to the molecule of water.

3) *The Durbin-Frank-Solomon's equation (1956).* These authors have related σ to the effective pore areas for filtration of water and solute as follows

$$1 - \sigma = \frac{A_s}{A_w}. \qquad 4\text{–}17$$

Combining equation 4–16 and 4–17 we obtain:

$$1 - \sigma =$$

$$\frac{\left[2\left(1 - \frac{a_s}{r}\right)^2 - \left(1 - \frac{a_s}{r}\right)^4\right]\left[1 - 2.104\left(\frac{a_s}{r}\right) + 2.09\left(\frac{a_s}{r}\right)^3 - 0.95\left(\frac{a_s}{r}\right)^5\right]}{\left[2\left(1 - \frac{a_w}{r}\right)^2 - \left(1 - \frac{a_w}{r}\right)^4\right]\left[1 - 2.104\left(\frac{a_w}{r}\right) + 2.09\left(\frac{a_w}{r}\right)^3 - 0.95\left(\frac{a_w}{r}\right)^5\right]} \qquad 4\text{–}18$$

that is, an equation which relates the ability of a given molecular species to develope an osmotic pressure across a membrane whose pores are of a size comparable to the size of the molecules.

Equation 4–18 was used by Goldstein and Solomon (1960) to generate the curves shown in figure 4–16. Each curve relates the radius of a molecule (*a*) with its reflection coefficient (σ) for a pore of a given radius (*r*). Next Goldstein and Solomon developed a null method to measure σ in red cells.

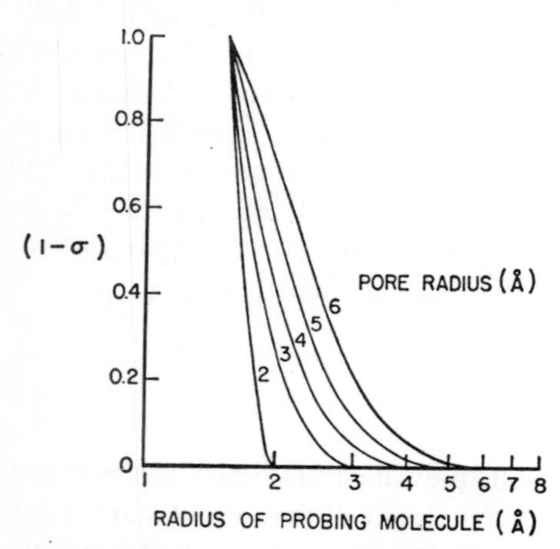

FIGURE 4–16 $(1 - \sigma)$ as function of the radius of the permeating molecule (Taken with kind permission from D. Goldstein and A. K. Solomon, 1960).

It depends upon the determination of the rate of swelling of red cells immersed in solutions of a probing molecule at several concentrations. If the probing molecule is smaller than the pore the cells will finally swell, regardless of the hypertonicity of the solution. However, if one uses a hypertonic solution whose concentration is sufficiently high, the cells will *initially* shrink. As the probing molecule is smaller and smaller, the necessary concentration to outbalance the swelling effect is higher and higher. For a given molecule, there is a concentration at which the swelling and shrinking tendency at zero time compensate each other and the volume of the cells does not change. What Goldstein and Solomon's method tries to measure is this equilibrium concentration $(C_s^o)_{\text{iso}}$. Using this parameter, and a formula whose basis were deduced by Kedem and Katchalsky (1958), they compute the reflection coefficient at zero time ($\sigma_{t=0}$). This formula, which only holds at $t = 0$ states that

$$\sigma_{t=0} = \frac{\Sigma_j C_j^i - \Sigma_j C_j^o}{(C_s^o)_{\text{iso}}}$$

in which $\Sigma_j C_j^i$ is the sum of the concentrations of all the non-permeant intracellular species and $\Sigma_j C_j^o$ the sum of the concentrations of all the non-permeant extracellular species. They used this procedure with a series of molecules ranging from urea (molecular radius 2.03 Å) to glycerol (molecular radius 2.74 Å) obtaining values of σ ranging from 0.58 to 0.88. Then they calculated which curve of the family given in figure 4–16 gives the best fit for the calculated set of σ. This curve is the one of a pore 4.2 Å in radius (fig. 4–17).

Methods of this type were used to measure the diameter and number of equivalent pores in a series of cells. See for instance Villegas and Barnola (1960), Curran (1960), Durbin (1960), Vargas and Johnson (1964). For an excellent review see Solomon (1968).

The size of the pores, as determined by these methods, seem to vary to adapt the membrane to different physiological conditions. Thus the axolemma of the squid axon in the resting state has a population of one pore of 4 to 5 Å of radius per 1.000 Å^2 (Villegas and Barnola, 1961; Villegas, Caputo and Villegas, 1962). Upon stimulation (100 impulses/sec) the radius changes to 6.2 Å (Villegas, Bruzual and Villegas, 1968). The antidiuretic hormone changes the pore radius at the outer surface of the epithelium of the toad skin from 4.5 to 6.5 Å (Whittembury, 1962). This effect was first discovered by Koefoed Johnson and Ussing (1953).

It is very important to keep in mind that the straight cylindrical pore described is an *equivalent* pore. As a model it might be as useful as the carrier model. As we shall see in chapter 6, it is of paramount importance in electrolyte permeation where not only the size, but also the charge of its walls have to be taken into account (see for instance Giebel and Passow, 1960; Pidot and Diamond, 1964).

Pores can discriminate molecules on the basis of size, shape and charge. The kinetics of penetration through pores also shows saturation. Two different substances using the same pore will also competitively inhibit the flux of each other. Pores do not show countertransport.

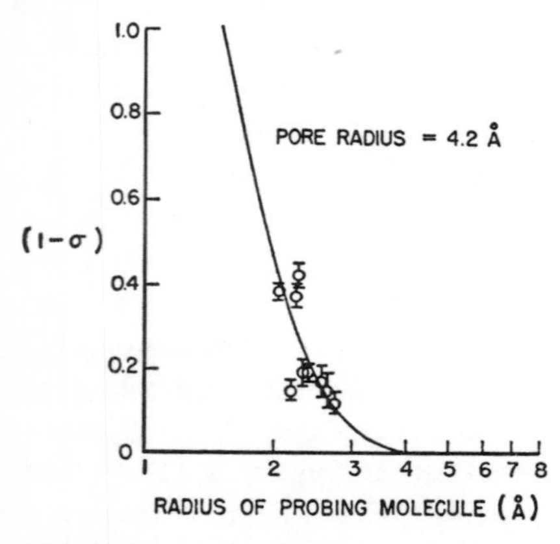

FIGURE 4–17 Experimental values of $(1 - \sigma)$ for several molecular species. The points are best fitted by a theoretical curve of 4.2 Å (Taken with kind permission from D.A. Goldstein and A.K. Solomon, 1960).

When the pores are narrow enough, so that molecules are forced to diffuse in single file, they show a very interesting phenomenon. Suppose a membrane with such a pore separates two solutions of different concentration of a given non-electrolite which can diffuse through the pore in single file only. A net flux from the high concentrated (side 1) to the low concentrated (side 2) solution will be observed. If we add a tracer of the substance to side 1 we can measure J_{12}; if we add it to side 2 we can measure J_{21}. The relationship between the two inidirectional fluxes will deviate from Ussing's ratio (1949b) for non-electrolytes

$$\frac{J_{12}}{J_{21}} = \frac{C_1}{C_2}$$

but will be given by

$$\frac{J_{12}}{J_{21}} = \left(\frac{C_1}{C_2}\right)^n$$

where *n* is the number of positions in the file (Hodgkin and Keynes, 1955). The effect arises from the fact that when tracer goes "upstream" moves against a net force. This phenomenon was observed with potassium in muscle (Hodgkin and Keynes, 1955) Chloride in muscle (Hodgkin and Horowicz, 1959; Adrian, 1961). See also Harris (1960), Sjodin (1965) and Hladky (1965).

4.9 Fixed sites

They are "stations" distributed usually across the membrane which can selectively adsorbe certain solutes. The facilitation of diffusion by this mechanism compares to the crossing of a brook by jumping from stone to stone. In this sense fixed sites are really *pores* with fixed sites. Diffusion through fixed sites gives rise to saturation, competitive inhibition and selectivity. The selectivity is not a function of the size of permeant molecule as with the pores, but it depends on the affinity as it was the case with the carriers. Although countertransport has been cited as convincing evidence in favour of mobile carriers, and against fixed sites, Essig, Kedem and Hill (1966) have introduced a model consisting of sites constrained to a row lattice (each row of sites is an independent transport unit), which, under certain circumstances can show countertransport. Fixed sites play a major role in ion permeation. For examples of this model see: Danielli (1954), Sjodin (1961), Hladky and Harris (1967). The limiting case is given when the sites are confined to the membrane-solution boundaries so that the solute is adsorbed before diffusion. The behaviour of such a model resembles in many respects the behaviour of carriers (Rosenberg and Wilbrandt, 1957).

4.10 General comment

In this chapter we have studied some of the mechanisms by which substances cross biological membranes. All these mechanisms have the common feature that the force which operates them arises as a consequence of an electrochemical potential gradient between both sides of the membrane and are therefore called *passive mechanisms*. Similar mechanisms can be found in non-living systems. In the next chapter we shall study *active mechanisms*. Active mechanisms are characterized by a movement of substance which cannot be accounted for by electrochemical potential gradients and require expenditure of metabolic energy.

Before we leave this subject it is worth emphasizing that, in order to simplify the explanations, we have considered only uncharged substances. However, the mechanisms described are also used by charged substances. When the substances are electrically charged, the use of these mechanism give rise to electrical phenomena. This subject will be considered in chapter 6.

5

Active transport

THE MECHANISM for the transfer of molecules across membranes which we have described so far, cannot explain the rate of permeation and steady state distribution of some of the physiologically most important substances. The mechanisms of simple or facilitated diffusion for instance, cannot account for the low concentration of Na^+ or high concentration of K^+ in the cytoplasm, and the uptake of some sugars and amino acids. However, it is not that the mechanisms themselves cannot account for these phenomena, but it is the driving forces which operate them which fail to explain for instance, how a substance can show a net flux against its concentration gradient. Since the forces arising from the electrochemical potential gradient cannot afford an explanation, it is assumed that the necessary energy to transfer the substances must come from the metabolism of the cell.

Jardetzky and Snell (1960, see also Jardetzky, 1960) have argued that according to Curie's theorem (1908) it is impossible for a force of a given tensorial order to be associated with a flow of a higher tensorial order. In other words: a chemical reaction (thought to be a scalar phenomenon) could not give rise to a directional flow of matter (a vectorial phenomenon). To visualize its meaning we may consider a water tank in whose center a chemical reaction starts. One cannot expect this reaction to proceed in a particular direction in the space. According to Jardetzky and Snell, the one and only way to effect a transport process by the scalar forces of the chemical reactions is to have a source and a sink set up by such reactions in distinct but continuous regions of space. They consider that the production of a constituent by a chemical reaction in one region of the space and its destruction in another region, gives rise to a thermodynamic vectorial force and a concomitant flow. However, Langeland (1961) and Morszynski, Hoshiko and Lindley (1963) have pointed out that Curie's principle does not forbid coupling between metabolic reactions and transport when it takes place in anisotropic media such as membranes. Besides, as pointed out by Mitchell (1963), enzymatic reactions are indeed vectorial at the molecular level. The fact that they are generally studied in a macroscopic system, where the molecules are oriented at random, leads one to believe that the reaction is a scalar phenomenon. In this connection, it does not surprise one that the coupling between fluxes and chemical reactions were evident in biological membranes where the participating molecules can be orderly oriented.

The use of metabolic energy to transfer substances across membranes is called *active transport*. The whole concept rests on the assumption that the membrane separates two rather diluted solutions so that the driving forces in most cases can be evaluated on the basis of concentrations and electrical potentials. If the substances inside the cytoplasm were not entirely contained in free solution, the electrochemical potential differences operating across the membrane could be quite different, making, perhaps, unnecessary to assume that the metabolism should afford some energy to translocate molecules. This consideration is made here because some authors consider that active transport constitutes the Achille's heel of the Cell Membrane Theory. These authors argue that if a substance were selectively absorbed to macromolecules inside the cytoplasm, it might accumulate in the cell and still have a low thermodynamical activity. However, one point of view does not exclude the other, *i.e.* absorption within the cytoplasm does not exclude active transport across the membrane. In chapter 7 we will analyse the case of epithelial membranes that, when mounted as a flat sheat between two lucite chambers filled with Ringer's solution, do transport substances at metabolic expense. In single cell membranes though, the experiments cannot be so clear cut. Nevertheless, for the sake of clarity, in the present chapter we will treat active transport from the view of the Membrane Theory and leave for chapter 9 the consideration of other possibilities. But the idea of how we can be sure that a substance is being actively transported will haunt us for the rest of the chapter. Thus if we consider that a substance is actively transported when it goes against its electrochemical potential gradient (Rosenberg, 1954)—even if we can overcome the difficulties to evaluate it—we have two major exceptions: a) *countertransport* (see page 102) and b) *downhill active transport*. This is the situation given when the neccesity of taking up a substance or eliminating a toxic, or a metabolic end-product is so great, or the time available to do it is so short (the case of some epithelial membranes like the kidney tubule) that the cell "does not wait" for diffusion forces to operate but it pumps the substance down the gradient to speed up the process.

Passive, unidirectional fluxes going in opposite direction through the same membrane depend on the chemical activity of the substance in the compartment where they come from (a_i). If, besides, there is no interaction between them, they observe the following relationship (Ussing, 1954).

$$\frac{J_{12}}{J_{21}} = \frac{a_2}{a_1} \qquad 5\text{–}1$$

and, in the case of ions

$$\frac{J_{12}}{J_{21}} = \frac{a_2}{a_1} \, e^{\frac{zF\Delta\psi}{RT}} . \qquad 5\text{–}2$$

If, because of other tests, one can insure that there is no interaction between opposite going fluxes, a deviation from these equations could be taken as indication of active transport. However we have already seen an exception of this rule when discussing single file diffusion (see page 110).

As we have seen in chapter 3, the flux of a substance is not only originated by its "classical" driving force, but to its coupling to the driving forces of the fluxes of other substances as well (see equations 3–18 and 3–20). We have also seen that the driving force associated to a chemical reaction arises from the affinity, which is a function of the free energy of the reaction (see page 69). Let us assume that a chemical reaction consumes oxygen and that the flux of oxygen (J_{ox}) is proportional to the variation of free energy of the reaction, then

$$J_{ox} \propto \Delta F.$$

In this way the progession of the chemical reaction could be followed through the oxygen consumption. If the same system has a solute flux (J_i), which depends on its electrochemical potential gradient ($\Delta\bar{\mu}_i$), according to the approach based on Thermodynamics of the Irreversible Processes fluxes J_{ox} and J_i will obey

$$\begin{aligned} J_i &= L_{11}\Delta\bar{\mu}_i + L_{12}\Delta F \\ J_{ox} &= L_{21}\Delta\bar{\mu}_i + L_{22}\Delta F . \end{aligned} \qquad 5\text{–}3$$

If L_{12} (or L_{21}) is not zero, part of the flux J_i will depend on the chemical reaction. Kedem (1961) proposes the existence of this non-zero coupling coefficient between the flux of i and the free energy change of the reaction as a criterion of active transport.

Another way of checking whether the flux of a substance depends on the energy supplied by the metabolism is to study the flux both, in the presence and in the abscence of metabolic inhibitors such as cyanide, iodoacetate, etc. But then we cannot be sure that the inhibitor does not modify the permeability of the membrane.

In view of the difficulties met in defining—and experimentally recognizing—active transport, Wilbrandt (1960) suggests to avoid the term active transport and use a more descriptive one: uphill or downhill transport. Danielli (1954) instead, suggests to keep active transport as an operational definition of a transient value.

Once we accept that the metabolic energy can be used to move substances we are faced with the problem of explaining how the coupling takes place. On the basis that most transporting systems behave like carriers operated by a Maxwell's demon, most active transport models envisage this mechanism (the pump) as a carrier in which one of the many equilibria is shifted at metabolic expenses. Thus Jacquez (1961) has examined the case of a membrane which transports and is also passively permeable to a substrate, and has derived equations for four different situations: a) metabolic energy is used to speed up the return of free carrier to the side from which the substrate is pumped. b) metabolic energy is used to accelerate the migration of the carrier-substrate complex in the direction of the active transport. c) metabolic energy is used to favour the carrier-substrate association at the side of the membrane in contact with the compartment where the flux comes from. d) metabolic energy is used to dissociate the carrier substrate complex at the opposite side. These models afford consistent explanations for initial uptake rates, competitive inhibitions and the effect of preloading. Experimental data are often insufficient to decide among the different models.

The change of affinity of the carrier at the boundaries of the membrane was first suggested by Franck and Meyer (1947) and is one of the most used model of pumps. Figure 5–1 taken from Solomon 1962 shows a carrier able to pump K^+ toward the cell. When the carrier is in the form X^-, it has high affinity for K^+ but on the cytoplasmal side, under the action of metabolism, it switches to a form (Y) with low affinity for K^+ (see also Shaw, 1954; Hokin and Hokin, 1959).

Rosenberg and Wilbrandt (1963) and Silverman and Goresky (1965) have derived theoretical models in which the carrier moves with the same easiness when it is complexed than when it is free, but in which metabolic energy is needed to change its affinity *on both* sides of the membrane. In this model the substance gets transported from the side where the carrier has the higher affinity, to the opposite side. One of the advantage of this model lies on the fact that it explains initial uptake rates, competitive inhibition and counter-transport, as well as steady state distribution.

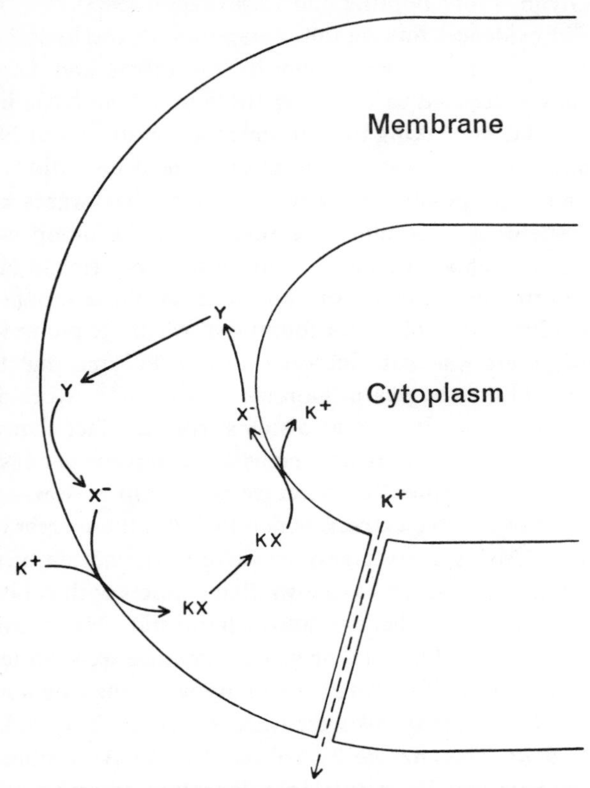

FIGURE 5–1 Schematic representation of a model in which a carrier with high affinity for potassium (X^-) is transformed into one with low affinity for potassium (Y). (Taken with kind permission from A. K. Solomon, 1962.)

5.1 Adenosinetriphosphatase (ATPase)

One of the best studied mechanisms of active transport is the one that transports Na^+. Studies carried out in giant axons have indicated that the extrusion of Na^+ is primarily coupled to the cellular metabolism and that the energy is afforded by the hydrolysis of the pyrophosphate bonds of the ATP molecule (Hodgkin and Keynes, 1955; Caldwell, 1956 and 1960; Caldwell, Hodgkin and Shaw, 1959; Caldwell, Hodgkin, Keynes and Shaw, 1960). The same relationship between Na^+ outflux and ATP splitting was observed in other biological preparations (Gardos, 1954; Dunham, 1957; Hoffman, 1960; Dunham and Glynn, 1961; Bonting and Caravaggio, 1963).

One of the best set of evidences linking ion transport with the hydrolysis of ATP comes from experiments carried out by Garrahan and Glynn (1967a, b) in which they succeeded in reversing the pump. The basic idea of these experiments is that, by making the concentration gradients of Na^+ and K^+ ions even more adverse than under physiological conditions, it should be thermodynamically possible to drive the pump backwards and synthesize ATP. The attempts to demonstrate reversal of the pump were made by suspending resealed ghosts rich in K^+ ions in solutions rich in Na^+ ions and then observing the rate of incorporation of intracellular inorganic phosphate labelled with P^{32} into ATP. They found that when the pump was being driven backwards, there was extra incorporation of P^{32} and that this extra incorporation was inhibited by pump-inhibitors like cardiac glycosides.

Interest in the active transport of Na^+ also stems from the fact that the movement of many other substances is associated with the movement of Na^+ (this point will be discussed in chapter 7). Therefore of great interest was the discovery by Skou (1957) of a nerve extract, which includes the membrane, and which exhibits an ATPase activity highly sensitive to the nature of the cationic environment. Further research has shown that extracts with ATPase activity can be obtained from many other tissues which transport Na^+actively (for references see Skou, 1965). Following the general practice we shall refer to this extract as "the enzyme" as if its ATPase activity were due to a single enzyme, even when it is dubious that this were the case (Hokin and Hokin, 1963a and b; Glynn, 1968). The enzyme hydrolizes ATP to ADP plus an inorganic phosphate and its activity is therefore expressed as micromoles of phosphate released per unit time and milligram of protein. In the presence of Mg^{++}, Na^+ and K^+ the extract increases considerably its activity. There is at present no conclussive evidence as to whether this effect is due to an increased activity of the "basal" ATPase or to the onset of a different enzyme. Since Skou's original discovery many other ion sensitive ATPases were found. Most of them have the same general behaviour but, of course, the ATPase extracted from the different tissues differ in some characteristics. For the sake of simplicity, unless otherwise stated, our description is based in the ATPase studied by Skou.

The enzyme requires Mg^{++} but this cation increases only slightly the activity. If, besides Mg^{++}, Na^{+} is also added, the ATPase activity increases further. This effect is quite specific for Na^{+}. The stimulation by Na^{+} is larger than that observed with other monovalent cations such as K^{+}, Rb^{+}, Cs^{+}, NH_4^{+}. However, if Mg^{++} and Na^{+} are present, the addition of K^{+} or Rb^{+}, or Li^{+}, or Cs^{+}, or NH_4^{+} produces a large increase of the ATPase activity (fig. 5–2). The value of Km indicates that, among this group of monovalent cations, K^{+} has the highest affinity for the enzyme. In the

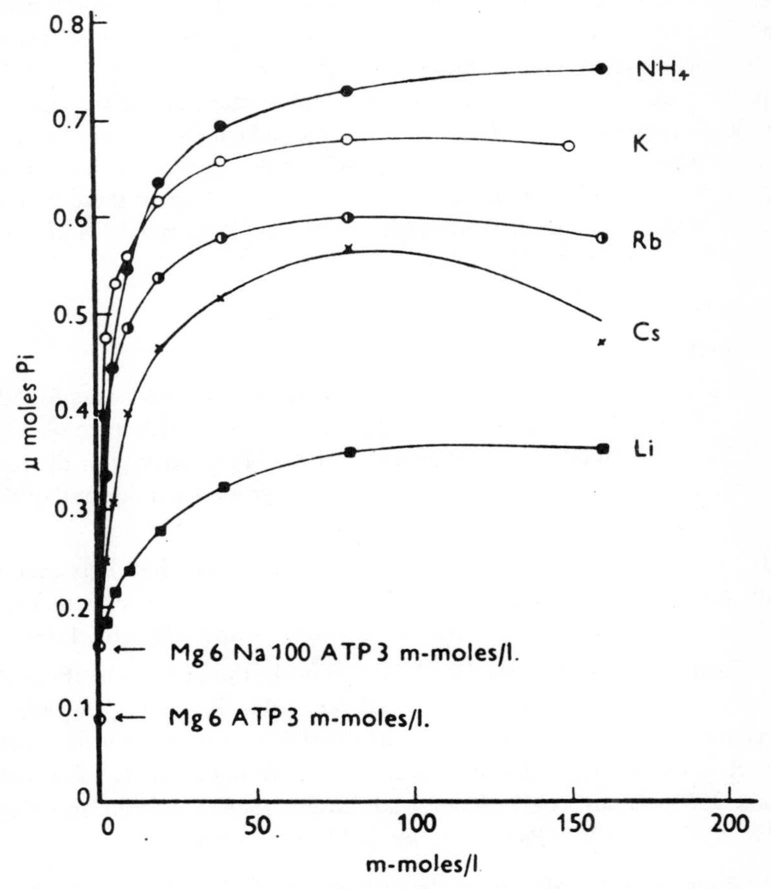

FIGURE 5–2 Effect of monovalent cations on the ATPase activity in the presence of Mg^{++} and Na^{+} (Taken with kind permission from J. C. Skow, 1963).

studies illustrated in figure 5–2 the concentration of Na^+ was kept constant at 100 mM. At lower Na^+ concentrations (let us say 20 to 40 mM), as the concentration of K^+ is raised, the ATPase is first stimulated by K^+ and then (beyond 20 to 30 mM K^+) it is inhibited down to the Mg^{++} level. The concentration of K^+ for maximal stimulation shifts toward higher values as the concentration of Na^+ is raised. This has two interpretations: a) the enzyme has a site (site *i*) which is 6 to 8 times more sensitive to Na^+ than to K^+, and another one (site *o*) which is highly sensitive to K^+ and has low Na^+ affinity. The maximal activation is achieved when site *i* is combined with Na^+ and site *o* with K^+. At all concentrations potassium occupies site *o*. If its concentration is much higher than the concentration of Na^+, K^+ also tends to occupy site *i*. Since this is not the ideal situation for the enzyme, its activity is reduced. b) There is only one site, which shifts from the outside to the inside of the membrane and possesses different affinities depending on which side of the membrane is. Regardless of which of these two possibilities is the real one, the properties of the ATPase resemble closely the properties of the pump. Thus:

a) both systems are localized in the membrane

b) both systems need ATP

c) The pump takes up Na^+ at the inside (side *i*) where it was found that K^+ competes with Na^+ and takes up K^+ at the outside (site *o*). On the other hand Glynn (1962) has shown that the ATPase activity of the red cell membrane requires the presence of Na^+ on the cytoplasmal side and of K^+ on the bathing solution.

d) The concentrations of Mg^{++}, Na^+, or K^+ giving half maximal activation are similar to those in which the active transport of cations has also half maximum activity (Post, Merrit, Kinsolving and Albright, 1960).

e) Cardiac glycosides and Ca^{++} inhibit both, the pump and the ATPase activity dependent on the presence of Na^+ plus K^+ ions. The necessary concentration of cardiac glycosides to affect both processes is also similar. In both cases the glycosides seem to act by interfering at the K^+ site (Schatzmann, 1953; Glynn, 1957; Koefoed-Johnsen, 1958; Caldwell and Keynes, 1959; Hoffman, 1962; Page, Goerke and Storm, 1964).

ATP increases the affinity of the enzyme for Na^+ and also the activating effect of K^+. The cyclic operation of the combination with ATP, its hydrolisis and subsequent release of ADP and P_i by the enzyme would also change cyclically its affinity for cations. This would make possible the operation of the enzyme as a pump. A scheme of how it would function was proposed by Skou (1965). TS represents the enzyme system. According to this scheme, in the deactivated enzyme, site *o* is occupied by Na^+ (c indicates that this Na^+, although it is on the outer side, comes from the cytoplasm). Site *i* is occupied by K^+, (b indicates that this K^+ was taken from the bathing solution). The number of ions transported per cycle is n, and is of the order of 2 or 3 ions (Sen and Post, 1961; Glynn, 1962; Baker, 1965; Whittam and Ager, 1965). Upon fixation of ATP, the affinity of the sites is reversed. The enzyme releases Na^+ at the site *o* and takes up K^+ while at site *i* it releases K^+ and takes up Na^+

$$\begin{array}{ccc} nNa^c + nK^b & & nK^b + nNa^c \\ \overset{o}{|} & & \overset{o}{|} \\ TS + Mg^{++} + ATP & \longrightarrow & Mg - ATP - TS \sim \\ \underset{i}{|} & & \underset{i}{|} \\ nK^b + nNa^c & & nNa^c + nK^b \end{array}$$

When K^+ is at side *o* and Na^+ at side *i* the enzyme hydrolyzes the ATP. At the same time sites *o* and *i* change their affinities allowing Na^c switch to site *o* and K^b to site *i*. The system if now in the condition of recommencing the whole cycle. This allows the transport system to translocate the cations.

$$\begin{array}{ccc} nK^b & & nNa^c \\ \overset{o}{|} & & \overset{o}{|} \\ Mg - ATP - TS & \longrightarrow & TS + Mg^{++} + ADP + P_i \\ \underset{i}{|} & & \underset{i}{|} \\ nNa^c & & nK^b \end{array}$$

Two of the major gaps in the identification of the Na–K–ATPase with the ion pump are the explanation of how the ATP fixation modifies the affinity for the cations, and how the cycle of ATP hydrolysis moves the cations across the membrane. Rega, Garrahan and Pouchan (1968) for instance, have demonstrated that the interaction of the p-nitrophenylphosphatase (which is related to the ATPase system) with K^+ and Na^+, is affected by the presence of ATP. In order to obtain more information Garranhan and Glynn (1967a, b, c and d) carried out a series of studies in which they impaired the work of the Na^+ pump either by raising the concentration of inorganic phosphate in the red cell (thus making difficult the conversion ATP → ADP + P_i), and by removing all the K^+ from the outer solution (thus preventing the Na^+ for K^+ translocation). They observed that under these conditions, ATP is practically not consumed and the pump becomes a passive sodium carrier with some of the implications and characteristics of carriers mechanisms that we have studied in chapter 4. The most likely interpretations of this phenomenon is that the carrier combines with ATP at the inner boundary takes up a Na^+ and moves it outwards; once there, a pump under normal conditions would had exchanged the Na^+ for a K^+ ion, but since K^+ was removed from the outer medium, the only thing the ATPase could do is to exchange it for another Na^+. The lack of K^+ also prevents the split of ATP, so that the ATPase remains with the Na^+ and the ATP and cannot operate in its usual cycle. The exchange of an internal Na^+ for one from the bathing solution can be detected as an exchange diffusion mechanism (see page 100) *i.e.* the outflux of Na^+ goes up as the concentration of Na^+ of the *outer* solution is increased.

A clue of how ATPase could work as a pump is based on the known ability of ATP for inducing electron transfer in mitochondria (Chance, 1961; Klingenberg and Schollmeyer, 1961) and on the demonstration that the electron distribution of a molecule can modify the field strength of its anionic groups and change its ion selectivity (Eisenman, 1961; Ling, 1962; see also chapters 6 and 9). It was postulated (Skou, 1963a, b and c) that the ATPase is associated with a pore with fixed negative charges arranged in such a way that the field strength varies as a consequence of the electron redistribution involved in ATP hydrolysis. Thus a site with low anionic field strength prefers K^+ while if the anionic field strength is raised, it will release K^+ and take up Na^+. Figure 5–3 illustrates a pore with 3 sites. Because of the position of the electron site 1 has high field strength and, therefore has Na^+ attached (see page 114). Upon fixation of ATP the electron is shifted to site 3. Site 1 looses its affinity for Na^+ and takes up K^+ from the outer bathing solution while site 3 raises its field strength and binds Na^+. As the electron flows from site 3 to site 2 and site 1 the peak of anionic strength goes outwards dragging Na^+ out.

The tissue extracts showing the Mg-dependent Na–K-activated ATPase activity are rather crude. A large number of efforts were made to purify the enzyme. Several attempts were made using: differential (Skou, 1957) or density-gradient centrifugation (Emmelot and Bos, 1962), treatment with deoxycholate (Schwartz, 1962; Auditore, 1962; Skou, 1962; Rendi and Uhr, 1964), aging the membrane preparation (Schwartz, 1962; Skou, 1962; Kinsolving, Post and Beaver, 1963; Hokin and Reosa, 1964), treatment with highly concentrated salt solutions such as NaI (Nakao, Nagano, Adachi and Nakao, 1963; Nakao, Tashima, Nagano and Nakao, 1965) and LiBr (Matsui and Schwartz, 1966). The general observation is that purification, no matter how delicate and carefull it may be, cannot proceed beyond certain point without lossing the cation-dependent activation. The ATPase activity of the red blood cell ghost is inactivated by organic solvents even when the concentrations are not enough to extract lipid from the ghost. This treatment weakens essential bonds between proteins and phospholipids and the ATPase system, being an integral part of the membrane, becomes disorganized (Roelofsen, Baadenhuysen and van Deenen, 1966). The enzyme activity seems to depend on a degree of organization which is well above the molecular level. Delicate treatments perhaps do not damage the enzyme

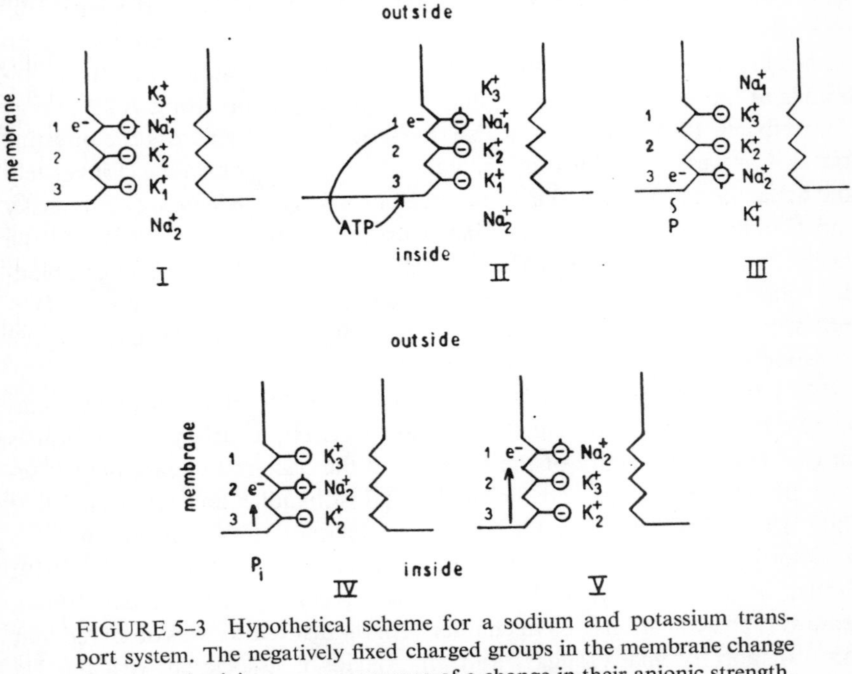

FIGURE 5–3 Hypothetical scheme for a sodium and potassium transport system. The negatively fixed charged groups in the membrane change their ion selectivity as a consequence of a change in their anionic strength, which is in turn induced by the electron transfer associated with ATP hydrolysis (Taken with kind permission from J.C. Skou, 1963b).

itself, but they might remove the necessary framework and the environment for its activity to be manifested.. In this sense, the purification of the pump looks like taking a radio and try to "purify" the loudspeaker: once it is "pure" we will not hear any sound. A promising approach is to try to attach the enzyme to an artificial membrane or other suitable solid support. An enzyme attached to a solid support may behave quite differently than in solution. It may change its optimum pH, sensitivity to ions and inhibitors, etc. (Dickman and Speyer, 1954; Goldman, Silman, Caplan, Kedem and Katchalsky, 1965; Racker, 1963; Brown, Chattopadyay and Patel, 1966; Whittam, Edwards, and Wheeler, 1968; Bobinski and Stein, 1966). For an interesting model of ATPase in which the cyclic change of the architecture of the membrane plays a major role see Opitz and Charnock (1965).

5.2 *Other pumps*

Although the transport of Na^+ and K^+ is the one in which the energetic, enzymatic and ion translocation aspects are more extensively studied, membranes are known to have other active transport processes. Thus the membranes of the sarcoplasmic reticulum transport calcium using ATP energy (Hasselbach, 1968). Calcium is also transported actively across the intestine and this process is modified by the addition of growth hormone (Finkelstein and Schachter, 1962). The skin of the frog *Leptodactylus ocellatus* (Zadunaisky and Candia, 1962) the rat intestine (Curran and Solomon, 1957) and the gastric mucosa (Durbin and Heinz, 1958) transport chloride. The gall bladder transports NaCl by means of an electrically neutral NaCl pump (Diamond, 1964). The thyroid gland transports iodide (Wolf, 1964). The stomach transports hydrogen (Heinz and Durbin, 1959).

The active transport of non-electrolytes is also observed in many membranes. Aminoacids, for instance, are transported by a variety of membranes ranging from the mucosa of the intestine to the bacterial membrane (Wiseman, 1951; Agar, Hird and Sidhu, 1953; Fridhandler and Quastel, 1955; Gale, 1953; Christensen and Riggs, 1952; Jacquez, 1961). Sugars are also transported by a variety of membranes (Crane and Mandelstam, 1959; Crane, Miller, and Bihler, 1961; Fox and Kennedy, 1965; Koch, 1964). However, as we shall discuss in chapter 7, in the last years it was observed that the transport of many sugars and aminoacids is associated with the Na movement and it seems probable that they are translocated through their coupling to Na pumping.

6

The movement and distribution of ions and their electrical consequences

DURING A long period which covers most of the first half of this century, knowledge on the theoretical relationship between fluxes and forces was, by far, more detailed than the knowledge on the actual conditions existing in the membrane to which the theoretical tool should be applied. Thus, in studying electrical phenomena under a variety of experimental conditions concerning concentrations, electric fields, partitions at the interface, etc., membranes were usually assumed to be homogeneous layers. The only departure from this approach was to assume that the membrane had fixed charges. The modern scope of the cell membrane, with the fixed or mobile sites; with ion-specific channels that can vary their permeability as a function of the electrical potential; with gates operated by one or more ions; with metastable states; with delicate couplings between metabolism and electrical phenomena; with its lipid component undergoing phase transitions, etc., offers a picture very rich in possibilities, but whose thorough treatment would require a detailed analysis of topic as different as the physical chemistry of glasses and the cation activation of ATPase, the dispersion and absorption in dielectrics and the chemistry of ring-shaped antibiotics. This, of course, escapes the purposes of this Introduction, where we consider preferable to give a general view of the problem. Therefore we chose for this chapter the following plan: first we will discuss briefly, and somewhat unconnectedly, the information that we consider essential to deal with electric potentials and electrolyte permeation in biological membranes: electroneutrality, diffusion potentials, the Nernst equation, Donnan distribution and potentials, ion exchange membranes, and some electrical phenomena associated with water movement. Then using this information, we will describe two equations that have played a major role in this field: the Goldman equation and the Hodgkin–Katz version of the Goldman equation. Once we have an idea of how ions move and distribute we will discuss how

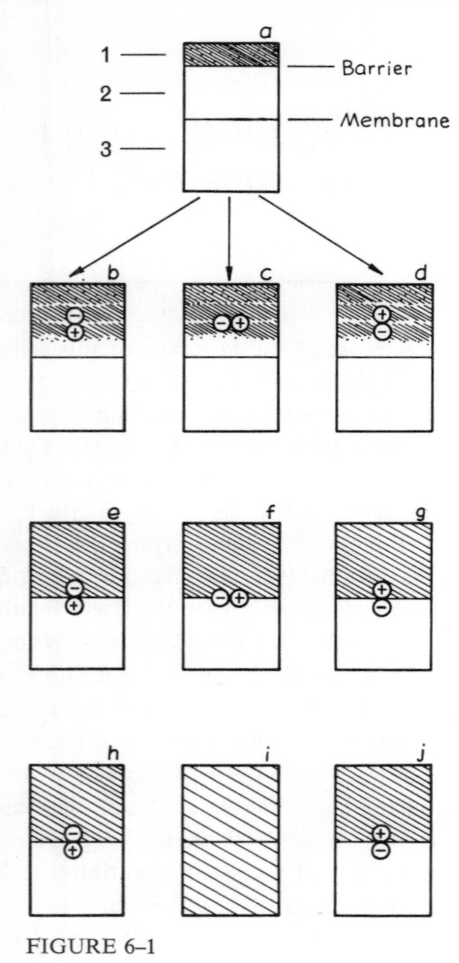

FIGURE 6–1

can the membrane tell the different ions apart (ion selectivity). We will later study a relatively new artificial system that gives information on carrier models: the interaction of macrocyclic antibiotics with lipid bilayers. At that point we will follow Eisenman–Sandblom–Walker and will classify the membranes on the basis of their structure and their basic mechanisms for ion translocation. This will help us to detect which mechanisms are operating in a given membrane. Finally we will illustrate the electrical implications of active transport.

Electroneutrality requires that the number of positive charges present in a given volume (the cytoplasm, the membrane, a mitochondrion, etc.) be equal to the number of negative charges. Separation of two unlike electric charges gives rise to an electrical potential that tends to bring them back together. In biological systems one finds quite a number of processes that provoke charge separation in a microscopic scale, and therefore, give rise to electrical potentials. Consider for instance the system depicted in figure 6–1. It consists of three compartments separated by a barrier and a membrane. All three compartments contain a solution of a mono-monovalent electrolyte BA, where B^+ is a monovalent cation and A^- a monovalent anion. The concentration in the upper compartment (1) is much higher than in the other two (2, 3). Upon removal of the barrier the electrolyte will diffuse down its concentration gradient. The flux of cation $B^+ (J_B)$ is described by equation 3–3. Considering that there is no gradient of hydrostatic pressure equation 3–3 becomes

$$J_B = -\omega_B C_B \left[\frac{RT}{C_B} \frac{dC_B}{dx} + zF \frac{d\psi}{dx} \right]. \tag{6–1}$$

The flux of anion (J_A) will obey the same equation. Macroscopically the concentrations of cation (C_B) and anion (C_A) have to be equal, otherwise the principle of electroneutrality would be violated. It also requires that J_B and J_A shall be equal. Hence

$$J_B = J_A. \tag{6–2}$$

Since equation 6–1 for the cation ($z = 1$) should be equal for the anions ($z = -1$) and using C instead of C_B or C_A

$$0 = -\omega_B \left[RT \frac{dC}{dx} + CF \frac{d\psi}{dx} \right] + \omega_A \left[RT \frac{dC}{dx} - CF \frac{d\psi}{dx} \right]$$

rearranging

$$\frac{d\psi}{dx} = \frac{RT}{F} \left(\frac{\omega_A - \omega_B}{\omega_A + \omega_B} \right) \frac{1}{C} \frac{dC}{dx}$$

$$\frac{d\psi}{dx} = \frac{RT}{F} \left(\frac{\omega_A - \omega_B}{\omega_A + \omega_B} \right) \frac{d \ln C}{dx}. \tag{6–3}$$

Notice that if the mobility of the cation (ω_B) is greater than the mobility of the anion (ω_A) (the case of H^+ and Cl^-), the cation will move ahead of the anion (fig. 6–1b) and the bottom of the system would be positive with respect to the top. It should be emphasized that although the mobilities may be completely different, the *velocities* should be the same, because the electrical potential developed affects the individual ion fluxes and enforces their compensation. If they have the same mobility (the case of K^+ and Cl^-), equation 6–3 predicts that there would be no electrical potential difference (fig. 6–1c). Figure 6–1d shows the case of a cation slower than the anion (the case of Na^+ and Cl^-). In this case the bottom will be negative. Once they reach the membrane, they will be retarded and, no matter how they moved in the free solution, their mobilities in the membrane (ω^m) will be different. The faster ion in the solution can still be the relatively faster in the membrane, but it might just as well be the other way around, so situations *b*, *c* and *d* could change to any one of *e*, *f* and *g*. The electrical potential in each case will follow the rules stated above, this time with the mobilities in the membrane instead of mobilities in the solution. Finally, if the membrane is permeable to both ions, the concentration would reach the same value everywhere in the system $\left(\frac{d \ln C}{dx} = 0\right)$ and the electrical potential will vanish (fig. 6–1i). If the membrane is impermeable to both ions, the concentration would reach an homogeneous value in the upper half, the bottom chamber will remain with its initial concentration, and the electrical potential will vanish. If only one of the ions is impermeable (say $\omega_A^m = 0$) equation 6–3 becomes (fig. 6–1h)

$$\frac{d\psi}{dx} = -\frac{RT}{F}\frac{d \ln C}{dx}$$

and integrating between limits yields

$$\Delta\psi = \psi_3 - \psi_2 = -\frac{RT}{F}\ln\frac{C_3}{C_2} \qquad 6\text{–}4$$

where C_3 and C_2 are the concentrations in chambers 3 and 2 respectively. This is the Nernst equation expressed in concentrations instead of activities. Notice that although the membrane is permeable to the cations, the electrical potential "compromises" with the concentration gradient and the ions do not achieve an homogeneous distribution. The same reasoning applies when ω_B^M is equal to zero, this time though, the sign of the potential changes (fig. 6–1j). Equation 6–4 widely used in membrane research to check whether the concentration gradient of a given ion agrees with the electrical potential. If it does not agree, it is assumed that energy is being spent to displace the equilibrium. This is often taken as a proof that the ion in question is being actively transported.

Another common situation of unequal distribution, this time involving a permeable anion *and* a permeable cation is the Donnan distribution. The additional requirement is the presence of a non-diffusible ion. This is the case of a system of two compartments separated by a membrane containing a solution of KCl, and an impermeable macromolecule with n negative charges ($P^{(n-)}$) confined to compartment 2. In this compartment electroneutrality requires that the concentration of positive charges (given by K^+) be equal to the concentration of negative charges (given by Cl^- and $P^{(n-)}$). Compartment 1 will therefore have

$$K_1 = Cl_1. \qquad 6\text{–}5$$

Compartment 2 will have

$$K_2 = Cl_2 + nP^{(n-)}. \qquad 6\text{–}6$$

According to equation 6–4, if K^+ and Cl^- distribute "passively", they will meet the following requirements

$$\Delta\psi = -\frac{RT}{F}\ln\frac{K_2}{K_1}$$

$$\Delta\psi = \frac{RT}{F}\ln\frac{Cl_2}{Cl_1}. \qquad 6\text{–}7$$

R, T, F and the electrical potential are the same in both equations. It follows that

$$\frac{K_1}{K_2} = \frac{Cl_2}{Cl_1}. \qquad 6\text{–}8$$

Replacing Cl_1 in equation 6–8 by K_1 (see equation 6–5) and Cl_2 by $K_2 - nP^{(n-)}$ (see equation 6–6) and rearranging it becomes

$$0 = (K_2)^2 - K_2 nP^{(n-)} - (K_1)^2$$

then

$$K_2 = \frac{nP^{(n-)} + \sqrt{(nP^{(n-)})^2 + 4(K_1)^2}}{2}.$$

We can now introduce this value of the concentration of K^+ of compartment 2 into equation 6–7 and obtain

$$\Delta\psi = -\frac{RT}{F}\ln\frac{nP^{(n-)} + \sqrt{(nP^{(n-)})^2 + 4(K_1)^2}}{2K_1}. \qquad 6\text{–}10$$

This is the expression of the Donnan potential originated by the presence of the charges attached to the non-diffusible macromolecule $P^{(n-)}$. Notice that if $P^{(n-)}$ disappears (imagine for instance, that it is metabolized and the products are permeable) equation 6–10 predicts that there will be no electrical potential difference between compartments 1 and 2. Since the cytoplasm has indeed many non-diffusible ions equation 6–10 is quite useful. Notice also that if the concentration of K were considerable increased, the Donnan distribution would be masked.

The negative charges do not need to be fixed to a macromolecule inside the cytoplasm to originate a Donnan potential. As discussed in chapter 1, the membrane also posses fixed charges, and they originate Donnan distribution and potentials. They will in fact originate two: one between the membrane and the cytoplasm, and the other between the membrane and the outer bathing solution (Wilbrandt, 1935; Teorell, 1936 and 1953; Meyer and Bernfeld, 1946).

6.1 The membrane as an ion exchanger

As said above, some of the molecules constituting the membrane posses electric charges. Although most membrane constituents are in a continuous rechange, this seems to be too slow as to play any significant role in permeation, so we may assume that the charges are confined to the membrane. They are generally called *fixed charges* to convey the idea that they do not leave the membrane. However it does not mean that they cannot ramble in the matrix of the membrane constituting *mobile* charges. We shall call *fixed charges* all those charges confined to the membrane until we know enough about them to make a distinction between mobile and truly fixed (see page 150).

Each fixed charge has a corresponding *counterion* of a charge of opposite sign so as to keep the whole membrane neutral. At this point it is convenient to emphasize that concepts valid for macroscopic systems (electroneutrality, validity of the flux equation, etc.) might not hold in a membrane 100 Å thick and with 100 mV across (10^4 volt per centimeter!). However, following the general usage, we will assume that the population of sites is large enough so that fluctuations do not impair electroneutrality. Since it was demonstrated that more than five jumps interposed in ion migration insure the validity of macroscopic flux equations to within 5 per cent (Ciani, 1965) we will also keep using these equations. Counterions can be exchanged for other ions of the same sign present in the bathing solutions. Thus the membrane may act as an ion exchanger. Mobile ions enter into the framework of the membrane not only to act as counterions of a fixed charge but also as ions dissolved freely in the water trapped in the membrane. This time though, in order to preserve electroneutrality, the mobile ion should enter accompanied by another mobile ion of the opposite sign: the *coion*. When an ion exchanger has charges of a sign only (say negative), all possitive mobile ions are called *counterions* and all negative mobile ions are called *coions*. The ion exchanger thus contain more counterions than coions. Coions find it harder to cross the membrane because of the repulsion of the fixed charges. When an electrical potential is applied across the membrane, more counterions flow in one direction than coions in the opposite. The *transport number* (t), which is defined as the fraction of the electric current which is carried by species i, is larger for counterions than for coions. The momentum imparted to the water by counterions is larger than the one imparted by coions and water will flow in the direction of the counterion transfer *(electroosmosis)*. Conversely, a net movement of water brought about by a non-electric factor (for instance: a hyperosmolar solution of sucrose on one side of the membrane), may drag ions and originate an electrical potential *(streaming potential)*. Streaming potentials are often provoqued experimentally to see whether the membrane has charged pores, and which sign is their charge (see for instance Pidot and Diamond, 1964). Notice that, in order to elicit these electrokinetic phenomena (electroosmosis, streaming potentials, etc.), the ion exchanger must have pores filled with solvent. If the pore is too narrow the counterion might need to lose its hydration shell before being admitted into the exchanger. Of course in this case coions are totally excluded.

The diffusion coefficient of ions through ion exchangers is much lower than in water. Ions migrate through ion exchangers *a*) by diffusing in the water which fills the pores or interstices (Mackie and Meares, 1955); Meares, 1956) and *b*) by jumping from fixed charge to fixed charge (Jakubovic, Hills and Kitchener, 1958 and 1959). The first mechanism is possible when there are liquid pores, the second one requires overlapping of the electric fields of neighbour fixed charges (Verwey and Overbeek, 1948). The penetration of water may swell the ion exchanger, pull the fixed charges apart, and reduce the overlapping. The development of hydrostatic pressure in regions of the membrane may thus originate very peculiar effects.

When an ion exchanger loaded with a counterion (say K^+) is immersed in a solution containing another counterion (say Na^+), K^+ will migrate from the exchanger into the solution, and Na^+ from the solution into the exchanger. The exchanger and the solution will contain both K^+ and Na^+. The concentrations of K^+ and Na^+ in the ion exchanger are not necessarily proportional to their concentrations in the solution, because the fixed charges may have a marked preference for a given ion species. This is called *ion selectivity* and we will discuss it later (see page 140).

The number of fixed charges in an ion exchange membrane determines the minimal concentration of counterions. If the membrane is in contact with a solution too diluted, the difference in counterion concentration between the exchanger and the solution will be large, the Donnan potentials developed between the membrane and the solutions will be high and this will exclude coions *(Donnan exclusion)*. Since in this case coions cannot penetrate the membrane, they cannot cross to the other side. Electroneutrality requires that in this case counterions could not cross in a net amount to the other side either. In other words, the counterion enters the membrane only if another counterion of the membrane enters into the solution to replace it. If we add labelled ions to the solution we will observe a flux solution → membrane of counterions but not of coions. Consider now figure 6–2. It illustrates a study of Teorell (1953) in which he calculated the electrical potential difference between two solutions separated by a negatively charged membrane. Both solutions contain KCl or NaCl or LiCl or HCl. Changes in concentrations are made in such a way that solution 1 is always 10 times more concentrated than solution 2. On the right hand side the solutions are so dilluted that the coion (Cl^-) is totally excluded ($\omega_{Cl} = 0$). Under this condition equation 6–4 predicts

$$\Delta\psi = -\frac{RT}{F} \ln \frac{0.001}{0.01}$$

at the temperature used (293° absolute) this gives

$$\Delta\psi = -58 \log \tfrac{1}{10} = 58.$$

Cations in this case can cross the membrane only in exchange for a cation from the other solution. This of course, is not reflected in the electrical potential. As we go left in the abscissa of figure 6–2, the concentrations are higher, the Donnan exclusion is less efficient, ω_{Cl} is no longer negligible, and the electrical potential is lower. Finally, when the concentrations are very high (10^4 and 10^3 M), the restriction on the coion is negligible and the

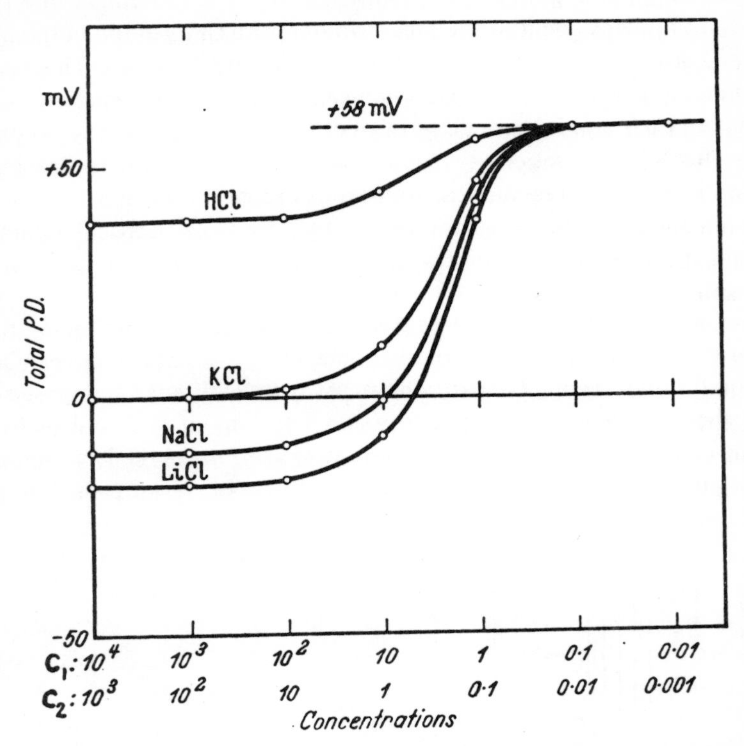

FIGURE 6–2 The total membrane potential across a negative membrane separating two solutions of a given electrolyte. The changes in concentration are made so as to keep always C_1 ten times higher than C_2 (Taken with kind permission from T. Teorell, 1953).

mobility ratio ω_+/ω_- is the same as in free solution. K^+ and Cl^- move with the same ease and do not originate an electrical potential H^+ moves faster than Cl^- and gives a positive potential. Li^+ is slower than Cl^- and they give a negative potential. For further insight in this kind of phenomena see Netter (1928), Meyer and Mark (1953), Sollner, Dray, Grim and Neihof (1954).

The study discussed above illustrates the profound influence that the charges of the membrane may have on its permeability. The red cell membrane, for instance, which has a net amount of fixed positive charges (Berg, Diamond and Marfey, 1965) is highly permeable to Cl^- and almost impermeable to Na^+. Figure 6–2 also illustrates that the movement of an ion depends on the movement of the other: an opposite charged ion in the same direction, or one of the same charge in the opposite direction. This ion of the same charge in the opposite direction can be of a different species, in this case the ion exchange is more evident. In cell membranes this may originate curious effects of ion selectivity (see for instance Passow, 1968). The interrelationship of the fluxes and distribution of different ions in a multi-ionic system like the cell membrane has important effects on its electrical potential. We will discuss further this point when considering the Goldman's equation (see page 136).

The interrelationship of ion fluxes gives rise to interesting, and biologically important effects even if the membrane does not posses fixed charges. Consider for instance figure 6–3. It consists of a "membrane" composed by several sheats of cellophane separating two solutions of HCl and NaCl of the same concentration. If we pass a current so as to make the left compartment negative, we help Na^+ to cross the membrane. If we pass it in the

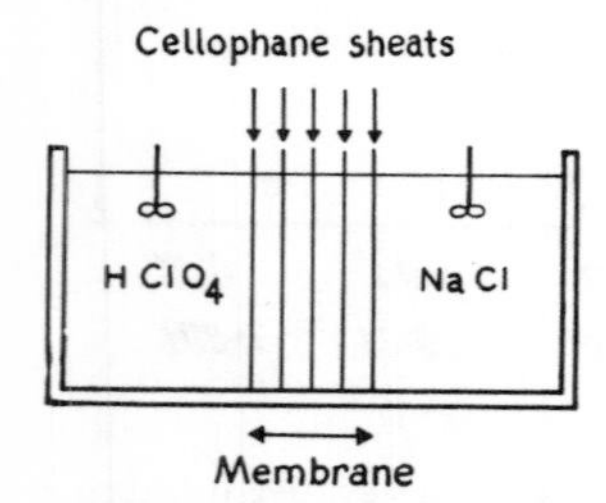

FIGURE 6–3 After T. Teorell (1956).

opposite direction we help H^+. We could check this chemically since this peculiar membrane permits to take samples and reconstruct its concentration profile (fig. 6–4). The difference between passing current in one direction or the other, is that the resistance of the membrane will be much higher when the current is carried through the membrane by Na^+ than when it is carried by H^+. This assymetry in the resistance (rectification) is analogous to the one observed in some biological membrane. For further discussion read Teorell (1956, 1958, 1959).

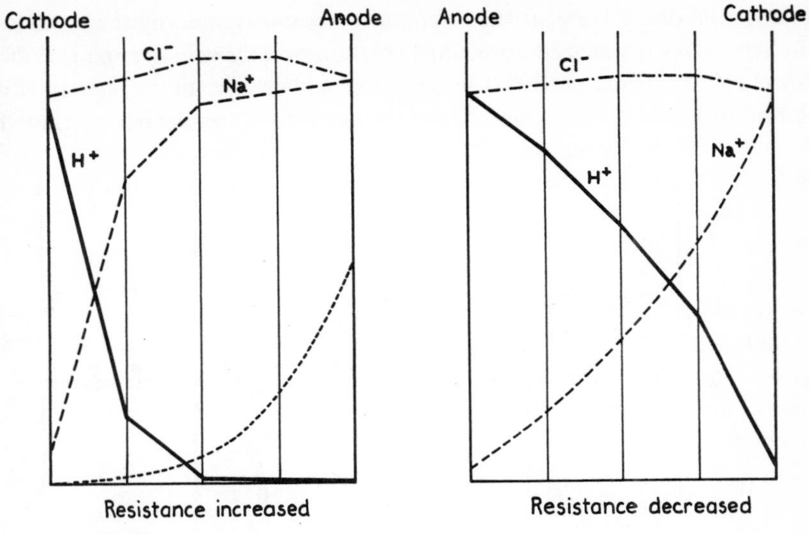

FIGURE 6–4 After T. Teorell (1956).

A general conclusion of what we have studied so far, is that the electrical potential difference between two compartments separated by a biological membrane consists, at least, of three different components

$$\begin{matrix}\text{Total} \\ \text{potential}\end{matrix} = \begin{matrix}\text{Donnan} \\ \text{potential}\end{matrix} + \begin{matrix}\text{Diffusion} \\ \text{potential}\end{matrix} + \begin{matrix}\text{Donnan} \\ \text{potential}\end{matrix}$$

This view neglects any role of the potentials that might arise as a consequence of the presence of unstirred layers (see Helfferich, 1962).

6.2 Some electrical phenomena associated to water movement

Consider figure 6–5. It depicts a pore whose wall has negative fixed charges. Electroneutrality requires that 8 positive counterions be present into the lumen. But the solution in the lumen has also 2 coions, which require two extra counterions. This makes a total of 10 counterions and 2 coions which conferes to the solution inside the pore 8 net positive charges. This shows that, although the whole pore of figure 6–5 (wall plus solution) is neutral, it is composed of a negative wall and a positive core. If water is made to flow—by applying, for instance, an osmotic gradient—ions will be displaced until the force of the dragg matches the electrical potential created by the separation of charges. As mentioned above this constitutes a *streaming potential.* If we cancel this electrical potential by applying a short circuit current, ions will be free to move as the water stream pushes them. The electric current that

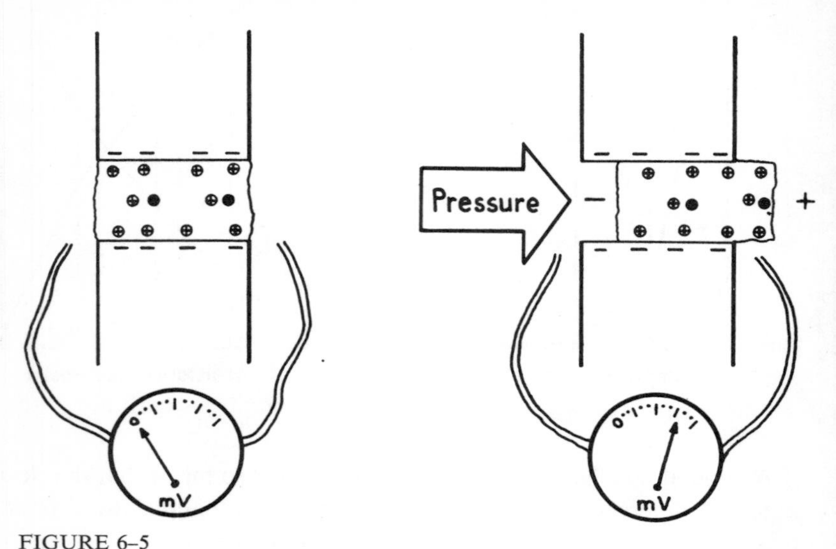

FIGURE 6–5

they carry is called *streaming current*. These phenomena are due to the viscous interaction between the ions and the molecules of water. Therefore it also works the other way around, *i.e.* by passing an electric current we may produce a water flow *(electroosmosis)*. In those biological membranes in which hydrostatic pressure is an important driving force (blood capillaries, renal glomerulae, etc.) these effects could play a role in ion movement and distribution.

6.3 The Goldman's equation as adapted by Hodgkin and Katz

As discussed in chapter 3, the Nernst–Planck equation has been integrated for a variety of special cases. One that has been particularly fruitful in membrane research is the Goldman's equation (1943). It would be convenient therefore, to discuss it briefly. It deals with membranes in which there is no gradient of pressure. The characteristic of Goldman's treatment is to assume

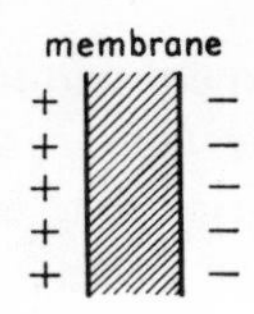

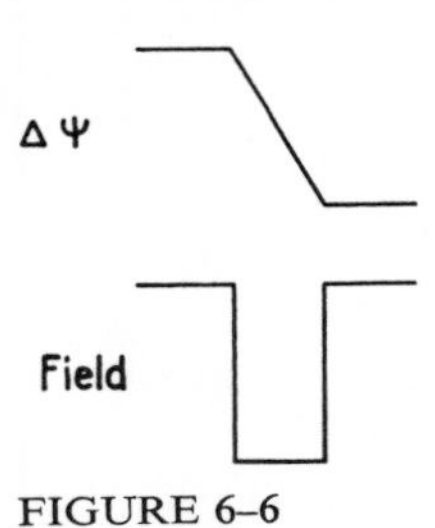

FIGURE 6–6

that the electric field across the membrane is constant, that is, $d\psi/dx$ is replaced by $\Delta\psi/\Delta x$ (fig. 6–6) and that there is no difference in hydrostatic pressure ($dP/dx = 0$). Equation 3–3 under these conditions becomes

$$J = -\omega RTC\left[\frac{1}{C}\frac{dC}{dx} + \frac{zF}{RT}\frac{\Delta\psi}{\Delta x}\right] \qquad 6\text{–}11$$

rearranging

$$-J\frac{1}{\omega RT} = \frac{dC}{dx} + \frac{CzF\,\Delta\psi}{RT\,\Delta x}$$

which is of the form

$$ib = \frac{dC}{dx} + aC \qquad 6\text{–}12$$

where

$$i = -J \quad b = \frac{1}{\omega RT}$$

$$a = \frac{zF\,\Delta\psi}{RT\,\Delta x}.$$

Integrating equation 6–12 we obtain

$$C = \alpha\,e^{-ax} + \frac{ib}{a} \qquad 6\text{–}13$$

where α is an integration constant.

At $x = 0$ the concentration C is equal to C_1 and equation 6–13 gives

$$C_1 = \alpha + \frac{ib}{a}. \qquad 6\text{–}14$$

Solving equation 6–14 for α and introducing this into equation 6–13 we obtain

$$C = \left(C_1 - \frac{ib}{a}\right) e^{-ax} + \frac{ib}{a}. \qquad 6\text{–}15$$

At the opposite side of the membrane ($x = \Delta x$) the concentration C is equal to C_2 and equation 6–15 becomes

$$C_2 = \left(C_1 - \frac{ib}{a}\right) e^{-a\Delta x} + \frac{ib}{a}$$

solving for i we obtain

$$i = \frac{a}{b}\left(\frac{C_2 - C_1 e^{-a\Delta x}}{1 - e^{-a\Delta x}}\right).$$

Now we replace the values of i, b and a and obtain

$$-J = \frac{zF\Delta\psi}{RT\Delta x}\,\omega RT\left(\frac{C_2 - C_1 e^{-\frac{zF\Delta\psi}{RT\Delta x}\Delta x}}{1 - e^{-\frac{zF\Delta\psi}{RT\Delta x}\Delta x}}\right). \qquad 6\text{–}16$$

If instead of using the general molar mobility (ω), we use the electrical mobility u (in $cm^2\ volt^{-1}\ sec^{-1}$) defined as $u = z\omega F$, equation 6–16 then gives the electric current instead of the ion flux. For monovalent cations ($z = 1$) this equation is

$$I_+ = -\frac{u_+ F\Delta\psi}{\Delta x}\left(\frac{C_2 - C_1 e^{-\frac{F\Delta\psi}{RT}}}{1 - e^{-\frac{F\Delta\psi}{RT}}}\right) \qquad 6\text{–}17$$

where u_+ indicates the mobility of the cation. Equation 6–16 for the current carried by a monovalent anion ($z = -1$) is

$$I_- = -\frac{u_- F\Delta\psi}{\Delta x}\left(\frac{C_2 - C_1 e^{\frac{F\Delta\psi}{RT}}}{1 - e^{\frac{F\Delta\psi}{RT}}}\right).$$

In this equation we can use the following transformation

$$\frac{e^{-x}}{e^{-x}}\left(\frac{a - b\,e^{x}}{1 - e^{x}}\right) = -\left(\frac{a\,e^{-x} - b}{1 - e^{-x}}\right)$$

and obtain

$$I_- = \frac{u_- F \Delta\psi}{\Delta x}\left(\frac{C_2\, e^{-\frac{F\Delta\psi}{RT}} - C_1}{1 - e^{-\frac{F\Delta\psi}{RT}}}\right). \qquad 6\text{–}18$$

Electroneutrality requires that, when no external current is passed through the membrane, the sum of I_+ (given by equation 6–17) plus I_- (given by equation 6–18) be equal to zero. Hence, equating 6–17 and 6–18 we obtain

$$(u_+C_1 + u_-C_2)\, e^{-\frac{F\Delta\psi}{RT}} = u_+C_2 + u_-C_1$$

taking logarithms

$$\Delta\psi = -\frac{RT}{F}\ln\frac{u_+C_2 + u_-C_1}{u_+C_1 + u_-C_2}$$

generalizing for all monovalent ions

$$\Delta\psi = \frac{RT}{F}\ln\frac{\Sigma_i u_{+i} C_2^{+i} + \Sigma_j u_{-j} C_1^{-j}}{\Sigma_j u_{+j} C_1^{+j} + \Sigma_i u_{-i} C_2^{-i}}. \qquad 6\text{–}19$$

This is Goldman's equation: the potential difference across a membrane is a function of the concentration and mobilities of the ions present. It should be emphasized that the concentrations, mobilities and potentials are referred to the membrane, not to the compartments separated by the membrane.

Hodgkin and Katz (1949) adapted this equation to the conditions in which the axon membrane works. First they made an assumption similar to the one we discussed in page 67, *i.e.* concentrations at the boundaries of the membrane are proportional to concentrations in the adjacent bathing solutions. The conversion factor is β. Then they defined permeability P as

$$P \equiv \frac{uRT\beta}{F\,\Delta x}.$$

Introducing P into equations 6–17 and 6–18 they become

$$I_+ = -\left(\frac{C_2 - C_1\, e^{-\frac{F\Delta\psi}{RT}}}{1 - e^{-\frac{F\Delta\psi}{RT}}}\right)\frac{F^2\,\Delta\psi P}{RT} \qquad 6\text{–}20$$

$$I_- = \left(\frac{C_2\, e^{-\frac{F\Delta\psi}{RT}} - C_1}{1 - e^{-\frac{F\Delta\psi}{RT}}}\right)\frac{F^2\,\Delta\psi P}{RT}. \qquad 6\text{–}21$$

Next they assumed that the only monovalent ions that play a significant role in membrane potential are K^+, Na^+ and Cl^-. Therefore, the total current across the membrane is

$$I = I_K + I_{Na} + I_{Cl} \qquad 6\text{–}22$$

where I_K, I_{Na} and I_{Cl} are the currents carried by K^+, Na^+ and Cl^-. Currents carried by K^+ and Na^+ are given by equation 6–20, and the current carried by Cl^- is given by equation 6–21. Introducing these expressions into equation 6–22 it becomes

$$I = \left(-P_K K_2 + P_K K_1 e^{-\frac{F\Delta\psi}{RT}} - P_{Na}Na_2 + P_{Na}Na_1 e^{-\frac{F\Delta\psi}{RT}} + P_{Cl}Cl_2 e^{-\frac{F\Delta\psi}{RT}} - P_{Cl}Cl_1\right) \times$$

$$\times \frac{F^2\Delta\psi}{RT\left(1 - e^{-\frac{\Delta\psi F}{RT}}\right)}. \qquad 6\text{–}23$$

Since no current is passed, the total current I is equal to zero. Hence redistribution gives

$$\Delta\psi = -\frac{RT}{F}\ln\left(\frac{P_K K_2 + P_{Na}Na_2 + P_{Cl}Cl_1}{P_K K_1 + P_{Na}Na_1 + P_{Cl}Cl_2}\right). \qquad 6\text{–}24$$

This is the Hodgkin–Katz version of the Goldman equation: the membrane potential is given as a function of the concentrations and permeabilities of the three most abundant electrolytes. The use of this equation made possible a considerable advance in the understanding of fundamental biological problems ranging from the electrical behaviour of a neuron to the concentration of the bile in the gall bladder (Weidmann, 1956; Hodgkin, 1957; Page, 1962; Diamond, 1962).

6.4 *Ion selectivity*

Biological membranes discriminate or react distinctively with different ions, which implicate that their pores, carriers, enzymes, surface polar groups, etc. "recognize" particular ion species. Consider, for instance, equation 6–24. If one uses a membrane in contact with solutions in which Cl^- is replaced by $SO_4^{=}$, since most biological membranes are negligibly permeable to this ion ($P_{SO_4} \simeq 0$), equation 6–24 becomes

$$\Delta\psi = -\frac{RT}{F}\ln\frac{P_K K_2 + P_{Na}Na_2}{P_K K_1 + P_{Na}Na_1}.$$

If we take P_K as reference ($P_K = 1$), this equation becomes

$$\Delta\psi = -\frac{RT}{F} \ln \frac{K_2 + \delta Na_2}{K_1 + \delta Na_1}; \quad \text{where} \quad \delta = \frac{P_{Na}}{P_K}. \qquad 6\text{–}25$$

In the axon membrane under this conditions (SO_4-Ringers) δ is around 0.01. This means that this membrane is about a hundred times more selective for K^+ than for Na^+. This selectivity, and its changes, make possible the function of the whole nervous system, which gives an idea of the biological importance of selectivity. In this section we will then review some information on the physicochemical basis of ion selectivity.

Electroneutrality suffices to explain how cations are told from anions. Size may explain why a small anion traverses a pore faster than a larger one. Density charge could account for the stronger attachment of divalent over monovalent cations. The problem arises when one tries to explain how ATPase can discriminate between K^+ and Rb^+; why a small change of the concentration of K^+ of the outer bathing solution varies the membrane potential and a similar change of the concentration of Na^+ does not; why a bacterial membrane can recognize and concentrate K^+ from "potassium free"* solutions, why an axon varies its sensitivity to cations in a few milliseconds, etc. The selectivity is generally expressed by the rank order in which a group of ions influences a given phenomenon. It is usually indicated as $A > B > C \ldots > Z$. Elements of the group 1A, alkali metal cations Li^+, Na^+, K^+, Rb^+, Cs^+, have received the lion's share of attention—particularly in biological studies—and, since there is more information available, we will base our description on these ions.

The orders of selectivity do not always follow an obvious pattern. They depart for instance, from the Hofmeister series of the hydrated radii of the ions, and are not related to their crystalographic radii (Höber, 1945). However, a hydrated ion consists of a core whose radius is approximately that of the crystalline compound, sorrounded by a cluster of bound water molecules in rapid exchange (Taube, 1962) and may have different degrees of hydration (Wiegner and Jenny, 1927; Jenny, 1932; Bungenberg de Jong, 1949). Lettvin, Pickard, McCulloch and Pitts (1964) suggested that the different monovalent cations can be told apart on the basis of their electric

* A sample of *Escherichia coli* incubated in a solution in which the flame photometer detects no K^+, may show a net uptake of K^+. This means that the membrane has a sensitivity which goes beyond the limit of the flame photometer, and has also a mechanism to accumulate K^+ to a concentration well in the range of the photometer.

fields at short distance. The most probable spacial configuration of the field near a poorly hydrated ion posses greater asymmetries and its temporal variations differ markedly from that near a well hydrated ion. The different hydrated cations can be also told apart on the basis of their steric properties because they will differ in their effective interaction radii, their most probable number of bound water molecules—hence the manner in which the hydrated ion can be deformed—and the probability of having instantaneously a given allowed packing.

The fixed anionic group also plays an important role. It is well known, for instance, that sulphonic resins can absorbe $K^+ > Na^+ > Li^+$ while carboxilic or phosphoric resins prefere $Li^+ > K^+ > Na^+$ (Gregor and Bregman, 1951; Bregman, 1954). Eisenman, Rudin and Casby (1957) suggested that the order of selectivity is governed by the *anionic field strength.* They reason that, in order to exchange an absorbed ion I^+ for an ion J^+ in the solution, ion I^+ has to detach itself from the fixed negative charge X^- and undergo hydration, and ion J^+ has to attach itself to the charge X^- according to the following equation:

$$IX + J^+ \rightleftharpoons JX + I^+ + \Delta G_{ij}^0. \qquad 6\text{–}26$$

ΔG_{ij}^0, the standard free energy change of the reaction consists of the following components:

$$\Delta G_{ij}^0 = (\bar{G}_i^{\text{hydr}} - \bar{G}_j^{\text{hydr}}) + (\bar{G}_j^{\text{fix}} - \bar{G}_i^{\text{fix}}) \qquad 6\text{–}27$$

where $(\bar{G}_i^{\text{hydr}} - \bar{G}_j^{\text{hydr}})$ represents the difference in partial molal free energies of hydration of the ions I^+ and J^+; and $(\bar{G}_j^{\text{fix}} - \bar{G}_i^{\text{fix}})$ represents the difference of their free energies of interaction with the fixed anion. The difference in hydration energy for a given pair of cations is a very well known quantity. The difference in interaction-energies for the same pair of cations depends on the *state* of the fixed charge X^- (how far apart is an X^- from the neighbour; how fixed or mobile they are; whether H_2O or other molecules also interact, etc.) and the *force field* of X^-. Therefore for a given state of X^-, its cation selectivity depends only on the force of its field. If the ion exchanger consists for instance of X^- groups spaced more than 5 Å apart and water is not admitted, then the free energies of interaction between the cation and X^- are given, in first approximation, by the internal energies as calculated by Coulomb's law

$$\bar{G}_i^{\text{fix}} \cong -\frac{332}{r_i^+ + r_x^-}; \quad \bar{G}_j^{\text{fix}} \cong -\frac{332}{r_j^+ + r_x^-} \qquad 6\text{–}28$$

where r_i, r_j and r_x are the radii of the mobile cations and the fixed anion X (for the calculation of the radius of the fixed anion see: Eisenman, 1962). Note that for a given pair of ions (fixed hydration energies, fixed cation radii) the energy of the interaction will only depend on the radius of the fixed anion: the larger the radius, the weaker the energy (see equation 6–28) and the looser the attachment. We may take an anionic site X^- of radius r_x^- and calculate the ΔG_{ij} of exchange between Cs^+ (taken as reference) and another cation (say K^+). Then repeat the calculation for the ΔG_{ij} between Cs^+ and some other cation (Rb^+, Na^+ or Li^+). Then do all over for another value of r_x^-. Figure 6–7 shows ΔU_{ij} as a function of r_x^-. When the values of the entropy change ΔS_{ij} are also known, values of ΔF_{ij} can be calculated from the ΔU_{ij} values. Each time the isotherm of an ion crosses the isotherm of another ion the order of selectivity changes. At around 0.9 Å for instance, the isotherm of Li^+ crosses the one of K^+. It means that on the left of 0.9 a system prefers Na > Li > K > Rb > Cs (the lower the ordinate position, the stronger the preference) while on the right the system preferes Na > K > Li > Rb > Cs. As we progress from the right to the left (low to high field strength) in figure 6–7 the sequences, indicated by the Roman Numerals I to XI are the following:

I	Cs > Rb > K > Na > Li	VII	K > Na > Rb > Li > Cs
II	Rb > Cs > K > Na > Li	VIII	Na > K > Rb > Li > Cs
III	K > Rb > Cs > Na > Li	IX	Na > K > Li > Rb > Cs
IV	K > Rb > Cs > Na > Li	X	Na > Li > K > Rb > Cs
V	K > Rb > Na > Cs > Li	XI	Li > Na > K > Rb > Cs
VI	K > Na > Rb > Cs > Li		

It is interesting that among the 120 possible combinations of the 5 cations (*i.e.* 5!) Eisenman's theory predicts that only 11 orders are possible. This prediction was checked in living and non-living systems and the agreement is most satisfactory. Non-living systems consist of electrodes made out of different kind of glasses, ion exchange resins, clays, etc. (Rothmund and Kornfeld, 1918; Jenny, 1927 and 1932; Schachtschabel, 1940; Krishnamoorthy and Overstreet, 1950; Eisenman, Rudin and Casby, 1957; Eisenman, 1961, 1962 and 1967, and Eisenman, Bates, Mattock and Friedman, 1965). The electrodes, for instance, are used to measure the electrical potential difference between two solutions of a salt of K^+, Li^+, Rb^+. Cs^+ or Na^+. The magnitude of the potential difference developed by each salt is a function of the sensivity of the glass. One can then order the cations of the salts on the basis of the electrical potential developed. This order characterizes the selectivity of the particular glass used. The familiar glass electrode to measure pH is an example of a material with high H^+ selectivity. Electrodes can be also prepared with other kind of glasses and, instead of measuring H^+ activity, measure for instance the activity of K^+, Na^+, etc. (Eisenman, Rudin and Casby, 1967).

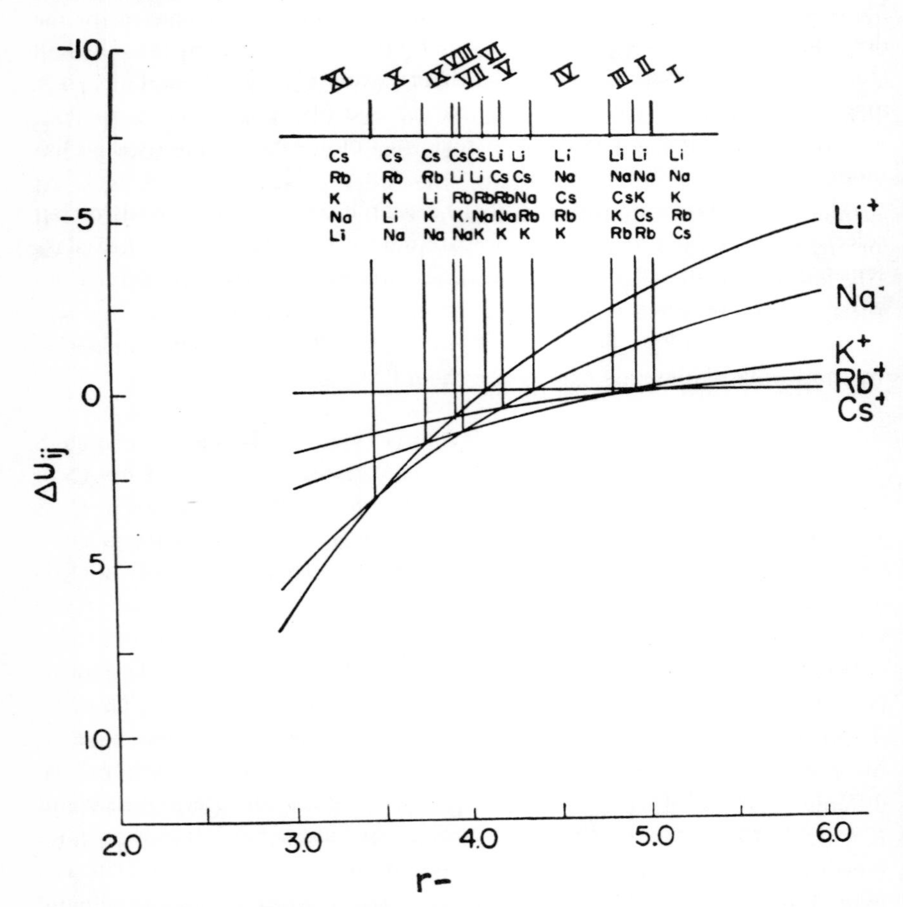

FIGURE 6–7 Cation selectivity expressed as ΔU_{ij} as a function of the anionic strength. The anionic strength decreases from right to left. When the values of the entropy change ΔS_{ij} are known, values of ΔF_{ij} can also be calculated (Taken with kind permission from Eisenman, 1961).

Biological data on selectivity consists of orders in which the different cations affect the electrical potential of the membrane (Osterhaut, 1939; Sjodin, 1959; Baker, Hodgkin and Shaw, 1962; Leb, Hoshiko, Lindley and Dugan, 1965) membrane conductances (Hodgkin, 1947), ionic fluxes (Cowie and Roberts, 1955; Conway and Duggan, 1958; Mullins, 1959; Sjodin, 1959; McConaghey and Maizels, 1962; Bolingbroke, Harris and Sjodin, 1961), nerve excitability (Cowan, 1934), ion exchange (Cohen, 1962), ATP hydrolysis by ATPases (Skow, 1960), etc. Although we have restricted the description to fixed, separated sites which do not admit water, the conclussions also hold for other states. The theory explains also the selectivity for anions exhibited by fixed possitive charges and the selectivity for divalent ions (Eisenman, 1965). For a recent review of biological selectivity see Diamond and Wright (1969).

As an important side product of the research on the atomic basis of ion selectivity, possible today it is to prepare microelectrodes made out of glass sensitive to a particular ion, impale a cell and measure the intracellular activity of the said ion (see Khuri, 1967).

6.5 Lipid bilayers

Bilayers of the type described in chapter 1 (see page 24) have a very high ohmic resistance (Mueller, Rudin, Tien and Wescott, 1964) and do not seem to possess for water permeation (Hanai, Haydon and Redwood, 1966; Cass and Finkelstein, 1967). However, this situation is completely changed upon the addition of certain substances which seem to become part of the artificial membrane and facilitate the diffusion of ions from one solution to the other. Thus the addition of "exitability inducing material" (EIM), a proteinaceous product of *Enterobacter cloacae* [possible a complex ribonucleoprotein (Kushnir, 1968)] lowers the resistance of the bilayer and conferes cation permeability (Mueller, Rudin, Tien and Wescott, 1962a and b; Mueller and Rudin, 1968a and b). If besides EIM, one adds protamine, the bilayer becomes also permeable to anions. If furthermore a salt gradient across is also added, the passage of a direct sustained current can produce action potentials similar to those observed in axon membrane (fig. 6–8). The observation that a small increase in the applied current elicits a large increase in conductance, suggests that EIM introduces into the bilayer a multivalent carrier or else, that it favours the micellization of the membrane. Micellization means that, when current is passed, part of the lipid of the artificial membrane changes reversibly from the bilayer to the micellar arrangement. Conduction through a membrane in which the lipids do not form a continuous hydrophobic diaphragm, is of course higher. It seems that the membrane pre-treated with EIM and protamine achieves a near-critical point and that when current is passed the membrane "flipps" back and forth from a state of cation to anion permeability, that results in the rhythmic firing

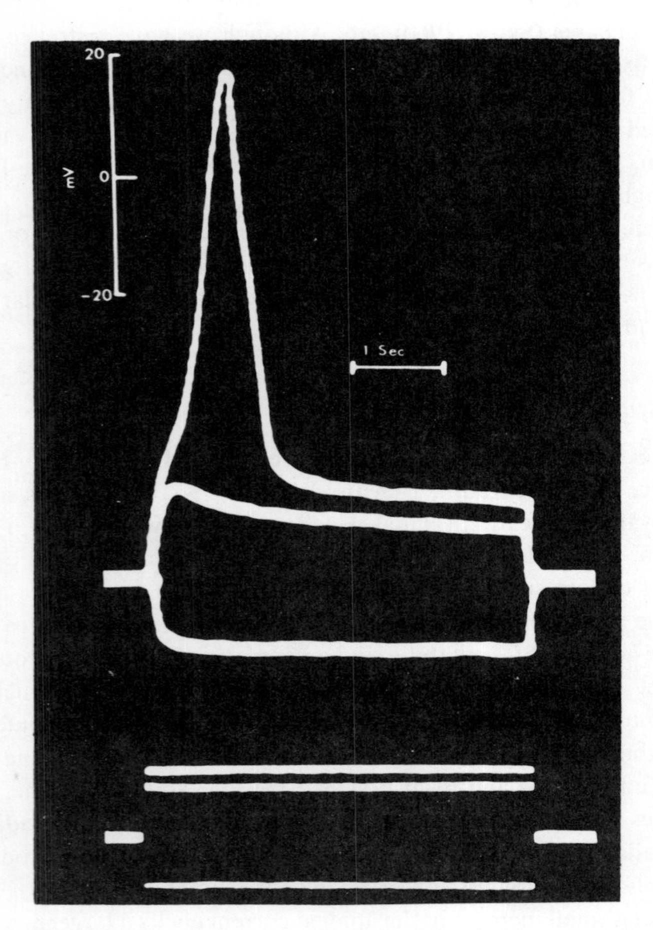

FIGURE 6–8 Action potential and subthreshold response in an experimental bimolecular lipid membrane (upper) in response to applied rectangular constant current (lower) (Taken with kind permission from P. Muller and D. O. Rudin, 1967).

of action potentials. In connection with the suggestion that micellization enables the membrane to fire action potentials, it is worth remembering the studies of Lucy, Glauert, Luzzati and others discussed in Chapter 1, in which they demonstrated that lipids may adopt in the membrane a micellar configuration. Some of these electrical effects can be also observed using alamethacin, a cyclopeptide antibiotic from *Trichoderma viride* (Mueller and Rudin, 1968b).

The interest in bilayer research has boomed since it was discovered that some antibiotics could be solubilized in the bilayer and act as carriers for the ions present in the bathing solutions (Lev and Buzhinski, 1967; Mueller and Rudin, 1967a and b). The beauty of these systems with mobile sites, is not only that they resemble closely the classical model of the membrane, but also that both, the physicochemical properties of the bilayer as well as the chemical structure of the antibiotic are known (Pedersen, 1967). It was later shown that synthetic cyclic polyethers, which are not antibiotic produce similar effects (Eisenman, Ciani and Szabo, 1968).

The common feature of these compounds is the ability to establish an ion-ion or an ion-dipole interaction with the mobile ion. In order to establish this interaction the antibiotic should be able to form a central focus of electronegative groups where the cation can be incased. There are three general forms in which this could be done (figure 6–9): a) the compound has the shape of a ring and the diameter of the orifice coincides roughly with the diameter of the mobile ion (*e.g.* valinomycin). b) the orifice is larger than the mobile ion, but the ion-compound interaction strains the ring and adapts the size of the orifice (*e.g.* the macrotetralide actins). c) the compound has a long shape (not a ring one) but the interaction with the ion wraps the molecule around the mobile ion (*e.g.* linear hexapeptides). However substances like the enniantin *B* have a small ring structure which is not large enough as to enclose a K^+ or a Na^+, yet they can increase cation per meation acros bilayers. The result of these forms of association is to hide hydrophobic groups (and the mobile ion) as much as possible, and leave the lipophobic part of the molecule in the perifery. The whole complex can now travel in the lipid component of the membrane. For a given compound, the ability of substituting for the hydration shell of the mobile ion, the constrain that the association imposes to the compound, the position of its hydrophobic parts after it has sequestered a particular ion, etc. depends on the nature of the mobile ion (see also Shemyakin, Ovchinnikov, Ivanov, Antonov, Shkrob, Mikhaleva, Estratoy and Malenkov, 1967).

Hence the process of ion translocation could show a considerable degree of cation selectivity. Most of the compound tested so far, if they have selectivity, they prefer K^+ over Na^+ (Lev and Buzhinski, 1967; Mueller and Rudin, 1967b; Andreoli, Tieffenberg and Tosteson, 1967; Tosteson, Andreoli, Tieffenberg and Cook, 1968). Some polyene antibiotics confere anion over cation selectivity and, among anions, the choice is made on size basis (Finkelstein and Cass, 1968).

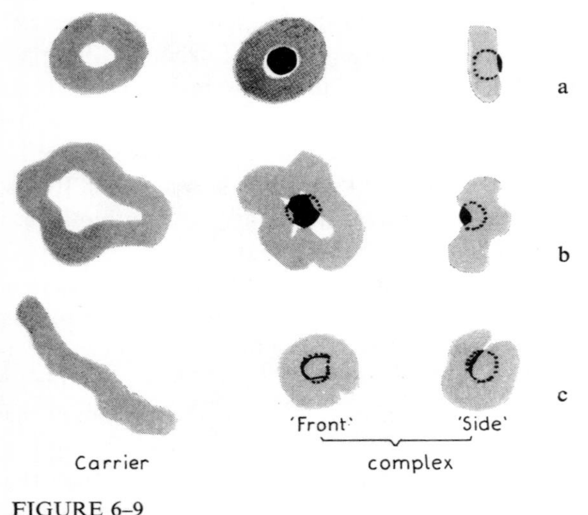

FIGURE 6–9

A very interesting alternative to the carrier model was proposed by Finkelstein and Cass (1968). They observed that while the increase in membrane conductance produced by compounds like valinomycin is *linearly* related to the concentration of valinomicyn, polyene antibiotics (Nystatin, Amphotericin B) increase the membrane conductance with the 6th to 10th power of the concentration. Finkelstein and Cass propose that polyene antibiotics form pores by attaching one another all the way across the membrane as depicted in figure 6–10. If this model is correct, some 10 molecules of antibiotic per pore would be needed. In agreement with the precarious stability that such a structure would have, Finkelstein and Cass observed that the increase in membrane conductance produced by the antibiotics is drastically reduced by small increases of the temperature. In those antibiotics that function as carriers it is the other way around. This ability to form pores is included here because it bears on bilayer permeability, however this mechanism has nothing to do with mobile sites. Besides, we have chosen bilayers to illustrate membranes with mobile sites because of their similarity with

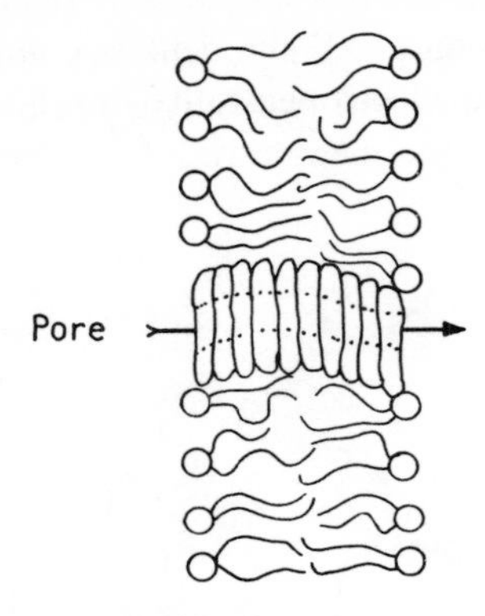

FIGURE 6–10

cell membranes, but they are just a particular kind of liquid ion exchangers. For other kind of liquid ion exchangers see Shean and Sollner (1966).

Eisenman, Szabo and Ciani (1968) (see also Szabo, Eisenman and Ciani (1969) carried out an illuminating theoretical and experimental analysis of

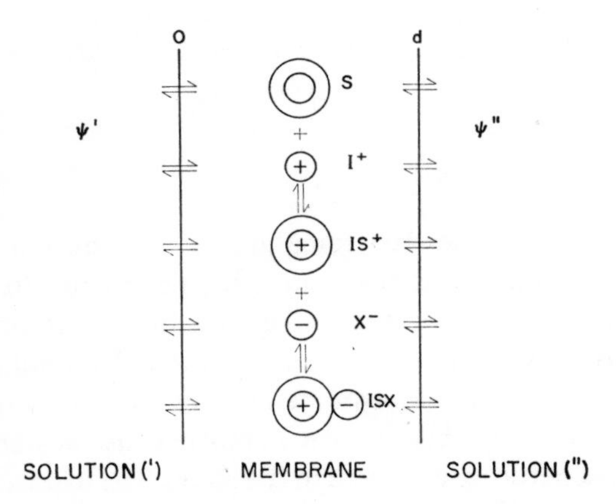

FIGURE 6–11 Diagram of the system used by G. Eisenman, S. M. Ciani, and G. Szabo (1968) to study the behaviour of the neutral molecular carrier of ions.

the role of neutral molecular carriers of ions in membrane phenomena. The model they adopted is depicted in figure 6–11. It consists of a lipid phase separating two aqueous solutions. S represents the neutral molecular carrier (monactin, see also figure 6–12), I^+ represents the free cation, IS^+ the complexed cation, X^- represents the free anion and ISX the neutral complex. Without using arbitrary assumptions as to electroneutrality, profiles of concentrations or electrical potential within the membrane, they correctly

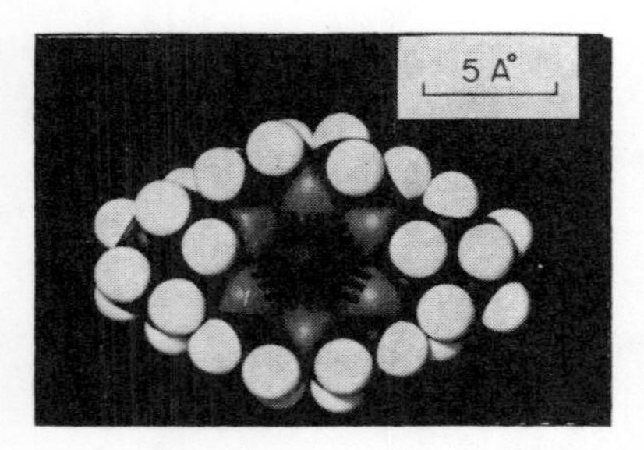

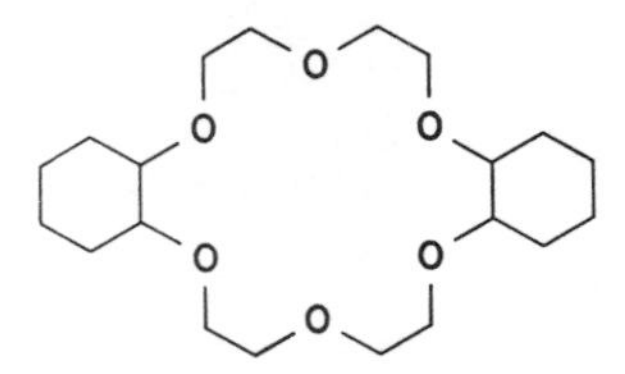

FIGURE 6–12 Chemical formula and space-filling model of the molecular carrier of ions monactin (Taken with kind permission from G. Eisenman, S. M. Ciani, G. Szabo, 1968).

predicted the behaviour of membrane potential and electrical resistance of phospholipid bilayer membranes, as well the ionic distribution equilibrium between aqueous solutions and organic solvents. The relative effects of Li^+, Na^+, K^+, Rb^+ and Cs^+ also support the view that monactin acts as a neutral carrier for ions.

6.6 *A classification of membranes*

In the previous sections we have described some general phenomena and basic mechanisms of ion permeation and distribution in membranes. In this section we will try to classify the membranes on the basis of their structure and the mechanism they use to operate the ion translocation. This will give us a group of criteria to characterize unknown membranes. We will restrict the discussion to homogeneous membranes. An homogeneous membrane is one which has the same properties everywhere in the plane. The distinction is important because membranes may have more than one mechanism working at the same time, either in parallel or in series, have regions with special properties, etc.

Following Eisenman, Sandblom and Walker (1967) we may classify homogeneous membranes in

- Homogeneous membranes
  - Without sites
  - With sites
    - Fixed
      - Dissociated FD
      - Associated FA
    - Mobile
      - Dissociated MD
      - Associated MA

6.7 *Membranes without sites*

If the central portion of the cell membrane were only constituted by the alkyl chains of the lipids, it would constitute a site-free membrane quite similar to a bilayer membrane of the type discussed above. The solubility of ions in a medium of low dielectric constant is so poor that it is doubtful that this route would play any biological role.

6.8 *Membranes with fixed sites—(solid ion exchangers)*

The factors which are important in this sort of membranes are: 1) the pore area (see also page 105). 2) The tortuosity of the pores. 3) The interaction ion-site. 4) The interaction ion-matrix (*i.e.*, not with the fixed charge but with the framework). 5) The electrokinetic phenomena. 6) The possibility that the chemical properties of the ion exchanger might change as a consequence of the ion exchange.

A. Membranes with fixed dissociated sites (FD) The important factors to be considered here are: 1, 2 and 5. This is the case of a solid ion exchange membrane with pores so wide that it allows bulk solution into the membrane. This offers the possibility of producing electrokinetic phenomena, which constitutes the distinctive characteristic of this kind of membranes. Coions are admitted on the basis of Donnan equilibrium and the mobilities of the different ions in the membrane phase are proportional to the mobilities in free solution. These membranes resemble quite closely some biological membranes. By adjusting the gradient of hydrostatic pressure, the current and the voltage, one may produce regions of negative resistance and self-sustained oscillations analogous to the firing of action potentials (Teorell, 1962).

B. Membranes with fixed associated sites (FA) The important factors to be taken into account here are: 3 and 4. Here the pore is so narrow that there is no water in free solution inside the membrane and the counterion is, by force, in close association with the fixed site. This makes the mobilities in the membrane much lower than in free solution. The fact that ions have to migrate in association with the fixed charges gives to these membranes high selectivity (Karrenman and Eisenman, 1962). The study of the mobilities and activation energies help to characterize the atomic mechanism of migration, in particular its dependence on temperature (see page 72) and on the mole fraction of mobile ions (Ilani, 1966; Eisenman, Sandblom and Walker, 1967). Since the only water present is that of hydration no electrokinetic phenomena can be elicited. Concentration profiles are formed when the membrane is in contact with two different solutions. Although the situation is different, the profiles are similar in several respects to those of figure 6–4. Since different counterions have different degree of hydration, these profiles may give rise to pressure gradients inside membrane. This, in turn, could stretch the matrix, pull the fixed charges apart and—since the ease of migration depends on the degree of overlapping of the fields of neighbour fixed charges—this may have profound effects on mobilities. Eisenman (1967) has shown that these effects may put the membrane in a metastable condition. Thus glass membranes separating NaCl and CsCl solutions could have a "state" in which resistance is high, and another in which it is low. The fact that this metastability is observed in a system like this, without any specialized chemical structure, might help to understand the basis of the excitation processes which are common in biological systems.

6.9 Membranes with mobile sites—(liquid ion exchangers)

The factor underlying all the differences between these membranes and those with fixed sites is the ability to rearrange the distribution of the sites according to the conditions present. Eisenman *et al.* have made an extensive and profound study comparing membranes with fixed and with mobile sites. A detailed analysis of these studies, however interesting, would take too long a digression and would go beyond the scope of this book. Therefore we will describe the main points and refer to the original papers for further readings (Conti and Eisenman, 1965a and b, 1966; Eisenman and Conti, 1965; Walker and Eisenman, 1966; Ciani and Gliozzi, 1968). This description is limited to those cases in which the coion is totally excluded. This condition is often met in experiments with biological membranes when sulphate or methylsulphate is used as anion. According to these studies the current-voltage relationship of membranes with fixed sites shows that, as voltage is increased, the current goes up and a *limiting conductance* is approached. In mobile site membranes, instead, as voltage is increased the current does not go up indefinitely but it approaches a *limiting current*. This effect is analogous to the existence of a J_{max} in the carrier mediated facilitated diffusion discussed in chapter 4. If the dissociation constant of the carrier-ion complex is too low, the membrane contains a large amount of ions in the form of neutral complexes. Eisenman has shown that in this case it takes a very high voltage to reach a saturation current. This voltage might be too high as to be used in biological preparations. Therefore, if we have a membrane that exhibits a rather linear current-voltage relationship, we cannot distinguish on the basis of the current-voltage relationship in steady state whether it is a fixed site or a mobile associated membrane. However, the distinction in this case can be made by comparing electrical and tracer measurements. In the mobile associated-site membrane, when the diffusion coefficient is measured on the basis of tracer fluxes, its value depends on *unidirectional* fluxes and has a high value. When it is, instead, calculated from the mobility (measured, in turn, by passing an electric current) on the basis of the equation:

$$D = \frac{uRT}{zF}$$

the value of D depends in this case on a *net* flux of current and has a low value. Therefore, even when membranes with fixed sites and membranes with mobile associated sites have a similar current-voltage relationship (at low voltages), a comparison of electric and tracer fluxes can tell one from the other.

As Eisenman, Sandblom and Walker (1967) pointed out, ionic mobilities in axon membranes are so low (of the order of 10^{-8} to 10^{-9} cm sec^{-1} volt^{-1} for K^+ and Na^+, Cole, 1965) that this membrane could not behave as a *FD* or *MD* but they could as *FA* and *MA*. However (see page 153), other biological membranes do show electrokinetic phenomena (see for instance Vargas, 1968), which would indicate that they could behave as *FD*. The existence of exchange diffusion of Na^+ (see page 100) also indicates that biological membranes possess *MA* systems.

Consideration of mosaic membranes (Neihof and Sollner, 1950; Mauro,

1962; Carr and Sollner, 1964) or discussion of dual or single ion channels (Finkelstein and Mauro, 1963) and the arrangement of membranes in series (Kedem and Katchalsky, 1963a, b and c; Teorell, 1953) will not be made here. However, the reading of the mentioned papers is strongly recommended.

6.10 Electrophysiological implications of active transport

So far we have studied electrical phenomena originated by the ion movement and distribution due to "passive" mechanisms. Biological membranes sometimes show electrical potentials which are difficult to explain on a passive basis, and which seem to be associated to the pump. As discussed above, the pump accumulated K^+ in the cells and extrudes Na^+, thus creating concentration gradients for these ions, and these gradients, in turn, originate diffusion potentials. Hence the pumping mechanism *is* at the bottom of some of the electrical potentials. But now we want to discuss a pump *directly* involved in the genesis of electrical phenomena: the so-called *electrogenic pump*. This situation seems to arise when the active translocation of ions is not neutral but it transfers a net charge.

Figure 6–13 shows the results of an experiment made by Page and Storm (1965) in which they studied the effect of temperature on the electrical potential and Na^+ and K^+ concentration of cat heart muscle *in vitro*. Cooling the muscles from 27.5 to 2°C reduces the metabolism and the active transport. The cells lose K^+ and gain Na^+. On rewarming to 27.5°C it takes about 40 minutes for the active transport to pump Na^+ out and K^+ in and restore the normal concentrations. The electrical potential across the membrane of these muscle cells depends on the ion distribution and diffusion (Page, 1962). Let us see then which are the electrical consequences of this operation. Figure 6–14 shows that muscles at low temperature (4.6°C) have a low potential. Rewarm produces an almost inmediate increase in the electrical potential (each dot is a value recorded with an intracellular microelectrode). It is clear that the electrical potential appears as soon as the pumping turns on, and *before* the ion concentrations are restored (some 40 minutes). This suggests that it is the pumping, not the ion distribution what accounts for the electrical potential, at least under this condition. Although this experiment could have alternative explanations, it illustrates the idea of an electrogenic pump. For further examples and rigorous discussion see: Bricker, Biber and Ussing 1963; Frazier and Leaf, 1963; Curran and Cereijido,

1965; Frumento, 1965; Mullins and Awad, 1965; Gonzalez, Shamoo, Wyssbrod, Solinger and Brodsky, 1967.

In this chapter we have described *single* mechanisms that the membrane may have to translocate ions and to produce electrical phenomena. The description was based mainly in non-living systems because very seldom a biological membrane has a unique mechanism. Cell membranes not only

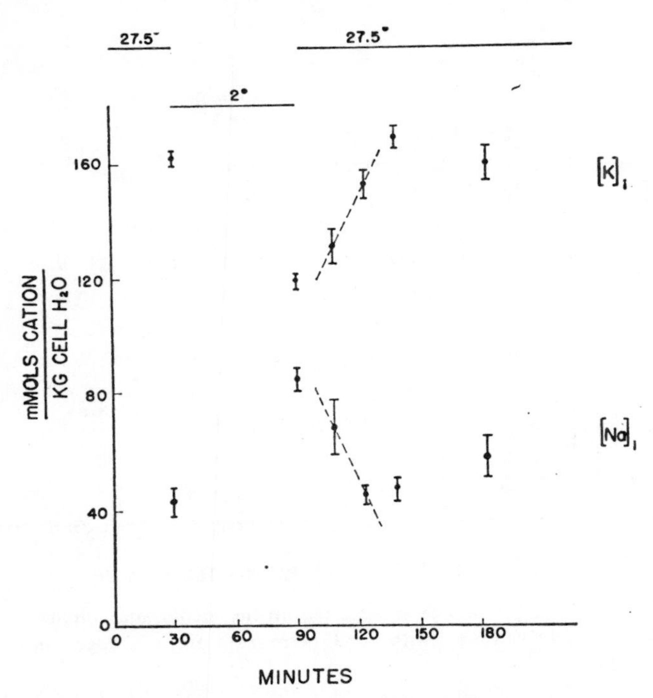

FIGURE 6–13 Effect of the cooling and re-warming on the intracellular concentrations of K^+ (upper plot) and Na^+ (lower plot) of muscle cells (Taken with kind permission from E. Page and S. Storm, 1965).

have translocations of many ion species going on at the same time, but could also translocate a given ion species by different mechanism. Na, for instance, is thought to cross red cell membranes and axon membranes using more than one kind of pumps, passive facilitated diffusion, simple diffusion, etc. (Hoffman, 1962 and 1966; Baker, 1968). None of them is a truly independent mechanism. They are related to each other and to the movement of all other ion species through electroneutrality, which gives an idea of the complicated business that the potential of a biological membrane is. Once we realize this, we marvel that in a given moment it should change in a strictly controlled way to generate and transmit the signal that makes possible the functioning of the nervous system and the contraction of muscles: the action potential. We will briefly refer to this phenomenon in chapter 8.

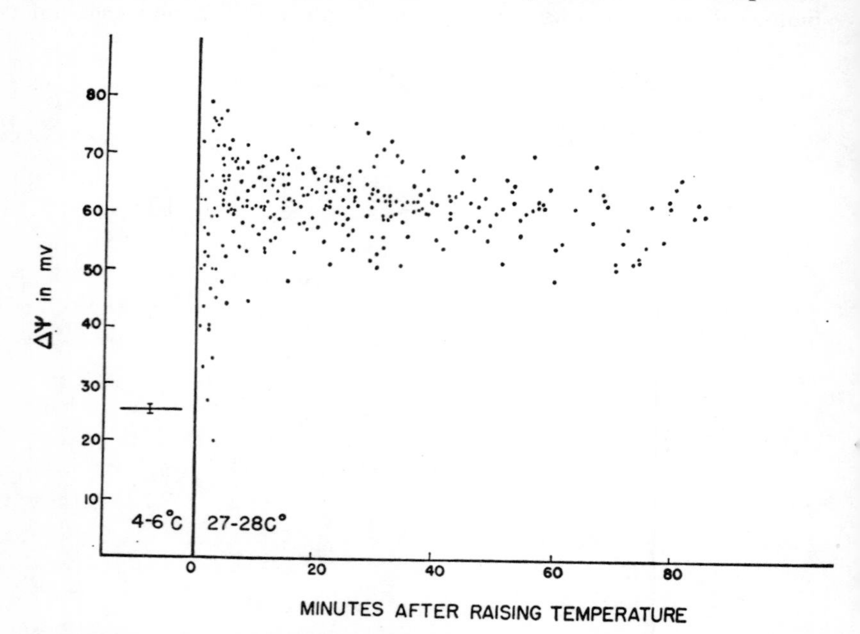

FIGURE 6–14 Effect of sudden re-warming on the membrane potential of muscle cells ($\Delta\Psi$) (Taken with kind permission from E. Page and S. Storm, 1965).

7

Epithelial membranes

THE EXCHANGE of substances between cells and extracellular solutions takes place at the level of the cell membrane, the exchange of substances between higher organisms and the environment takes place at the level of epithelial membranes constituted by one or more layers of cells. The intestinal mucosa, the renal tubes, the gall bladder and the frog skin are examples of epithelial membranes in which the whole epithelial cell—its membranes included—is interposed in the migration of substances. A substance may cross epithelial membranes by a) Penetrating through the outer facing membrane of an epithelial cell, diffusing through the cytoplasm, and crossing again the cell membrane at the opposite side of the cell (transcellular route). b) Passing between the cells. c) Going through specialized transporting compartments without getting mixed up with the pool of the same molecular species that may exist in the protoplasm. This could be done by travelling inside a pinosome (see chapter 8) or by migrating through membranous structures (see below).

In the preceding chapters we have considered membranes as if they were essentially identical everywhere around the cell. However a membrane can be for instance permeable to K^+ on one side of the cell, and impermeable on the opposite side. This striking asymmetry is typical of epithelial cells. Many of the most fundamental concepts in the movement and distribution of substances, in particular the relationship between fluxes and driving forces, come from the study of epithelial membrane. Not only because of their peculiar properties, but because with some of them one can control the solutions bathing both of their sides.

Figure 7–1 represents in a highly schematic way an epithelial membrane, the frog skin, in which we can observe that even the anatomy of the membrane is asymmetric. Among the characteristical features that are likely to have a functional role with respect to the translocation of substances, we may notice: a) The membrane has an outer part (pond side) occupied by the epithelial cells, and an inner part (body side) occupied by connective tissue. We shall keep the nomenclature of outer and inner side even in cases in which

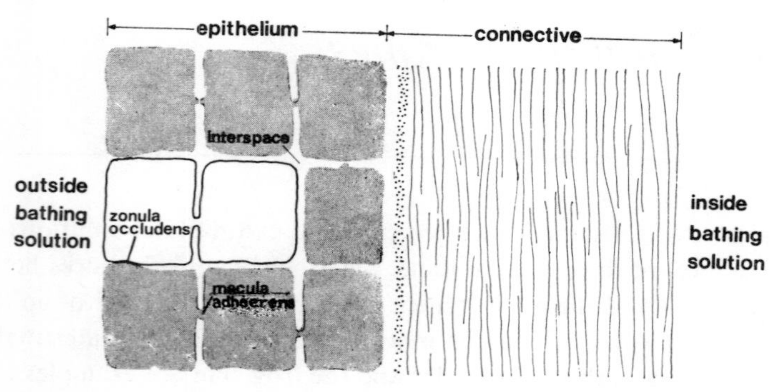

FIGURE 7–1 Schematic representation of an epithelial membrane, the frog skin. The blocks represent epithelial cells. Two cells in the center (not shaded) are represented in more detail in figure 7–10.

the membrane is dissected and mounted as a flat sheet between two Ringer's solutions as shown in figure 7–2. In the case of intestinal or gastric mucosa, gall or urinary bladder, renal tubules, etc., "outside" refers to the lumen or mucosal side, and "inside" refers to the interstitial or serosal side. b) In some epithelia not only the whole membrane, but also the individual cells are anatomically asymmetrical. Thus the position of the nucleus, the Golgi complex, the secretion granules, the mitochondria, the brush border, etc. are generally distributed in a definite pole of the epithelial cell (*see* Fawcet, 1962; De Robertis, Nowinsky and Saez, 1969). c) The cells are separated from each other by a narrow intercellular space which is closed toward the outside by *zonulae occludens*, and open toward the inside. The *zonulae occludens* are a special kind of junctions between neighbour cells. Another kind of cell junctions are the desmosomes or *maculae adhaerens*. A desmosome exists only at a discrete point, a molecule diffusing through the intercellular space can avoid a desmosome by detouring above or below the desmosomal level. The molecule could not sort the *zonulae occludens* by detouring, because these are continuous seals, all around the cells (Farquhar and Palade,

1963; Choi, 1963). The intercellular space contains a substance that seems to possess ion binding capacity (Farquhar and Palade, 1965). d) Kelly (1966) working with newt skin, has presented some evidence that the intercellular

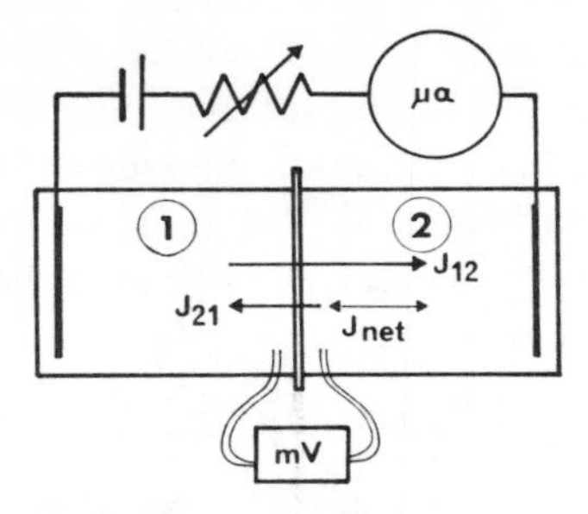

FIGURE 7–2 Experimental set-up for the study of epithelial membranes. The membrane is mounted as a flat sheat between two chambers (1 and 2) μa: microammeter. mV: voltmeter.

material of the desmosomes is arranged as pillars or partitions which are continuous with or layered upon the outer leaflet of the cell membrane as depicted in figure 7–3. Notice that in this way the outer leaflet of the cell membrane of the two cells would form a continuum. This arrangement of the outer leaflet may have a significant functional implication (see page 173). e) There is a continuous basal membrane between the epithelial cells and the

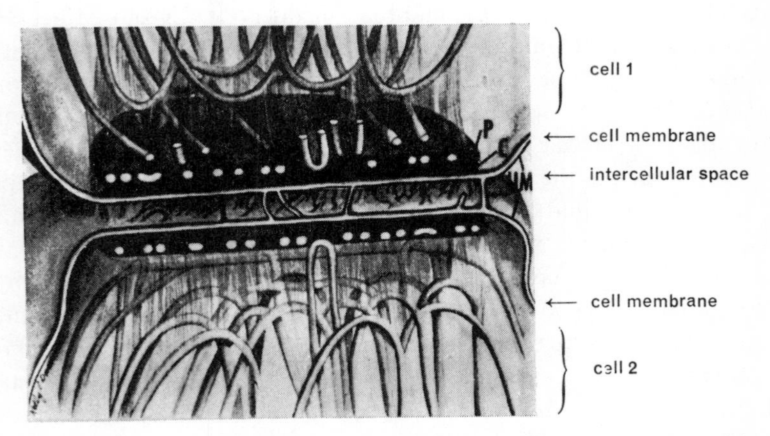

FIGURE 7–3 Schematic view of a large epidermal desmosome (Taken with kind permission from D.E. Kelly, 1966).

connective tissue represented in the figure 7–1 by a line of dots. f) Farquhar and Palade (1966) have demonstrated that the cellular borders facing the interspace contain ATPase (Fig. 7–9). Although the technique employed does not permit to discriminate between Na–K-ATPase and plain ATPase, this peculiar distribution, and in particular, its absence at the outer border of the epithelium might have, as we shall discuss later, an important functional consequence. g) Glands, *tela subcutanea*, and other anatomical characteristics of epithelial membrane do not appear to participate in the translocation of substances and are generally ommited from drawings made by biophysicists.

7.1 Cellular contacts and intercellular communication

Many cells in suspension tend to adopt a spherical form corresponding to a minimum surface energy. When they establish contacts with other cells, their shape and subsequent behaviour seem to be influenced by the membrane-membrane interaction. A variety of forces ranging from chemical bonds to triplets like $-COO^{-}Ca^{++-}COO-$ were suggested to participate in this association (see Weiss, 1958; Ambrose, 1964). In tissues like epithelial membranes, which are subject to special functional requirements cell-cell interactions also involve the formation of anatomical cell junctions of the type mentioned above (Campbell, 1967). We have also mentioned how the *zonulae occludens* seal the outer end of the intercellular space. This is precisely the site of the highest resistence to water flow in amphibian skin (MacRobbie and Ussing, 1961) and toad bladder (Leaf, 1960). Miller (1960) and Kaye, Pappas Donn and Mallett (1962) have demonstrated that the diffusion of macromolecules or particles do not go beyond the region of *zonulae occludens*. Muir and Peters (1962) suggested that these junctions would appear whenever a cell layer separates two compartments of different composition such as the tubular urine and the plasma; the water in the pond and the interstitial fluid of the frog. *Maculae adhaerens* offer only a localized restriction to diffusion with the only consequence of making the diffusion path a little more tortuous.

From what is said above, we may conclude that cell junctions constitute an element of restriction to the movement of substances *between the outer and the inner bathing solution*. On the contrary, they seem to constitute a low resistance path *between the cytoplasm of neighbor epithelial cells*. We will illustrate this point by discussing an experiment carried out by Loewenstein and Kanno (1964). (See figure 7–4). The upper part shows a salivary gland of the *Droso-*

phila with its giant cells represented as polygons. The cell on the left end is impaled with a microelectrode and a square pulse of current is passed between this microelectrode and the bathing solution (lower inset, upper trace). Another microelectrode records the resulting resistive membrane voltage across the membranes of the cells located at varying distances from the current-passing microelectrode. The lower inset (lower trace) shows the voltage recorded at 0, 200, and 800 μ. Several points of this experiment interest us here: a) As evidenced by the voltage elicited, the surface of the cells has a high resistance (about $10^4 \Omega . cm^{-2}$). b) It would appear that, in order to go from one cell to the next, current should traverse *two* of these resistances. c) However, the slope of the voltage-distance curve indicates that in going from cell to cell the current meets a much lower resistance (about four orders of magnitude lower) than the one that two cell membranes in series would offer. There must be, therefore, a connection of high conductance between the cells. d) This high conductance connection should be

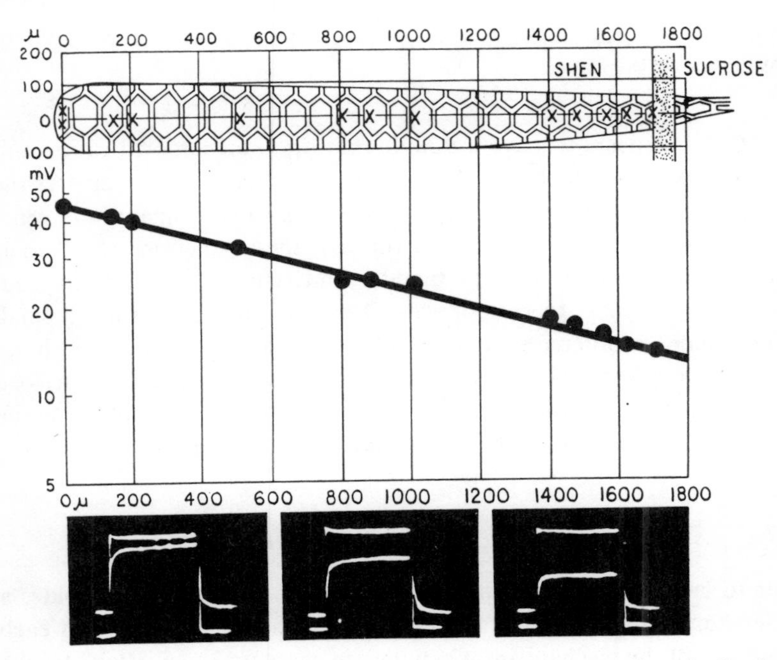

FIGURE 7–4 Membrane voltage attenuation along a series of cells. See text. (Taken with kind permission from W. R. Loewenstein and Y. Kanno, 1964)

insulated from the extracellular space, otherwise the current would have leaked to the bathing solution. Detailed studies carried out by Kanno and Loewenstein, (1964) (see also Loewenstein and Kanno, 1964 and Wiener, Spiro and Loewenstein, 1964) pinpointed the low resistance path at the level of the anatomical cells junctions discussed above. Figure 7–5 depicts one of these *junctional units* consisting of three principal elements (Loewenstein, 1966): 1) the junctional membranes *c*; 2) the surface diffusion barrier *s* and 3) an element of structural rigidity and adhesion *A*. This intercommunication between cells is permeable to fairly large molecules (molecular weight around

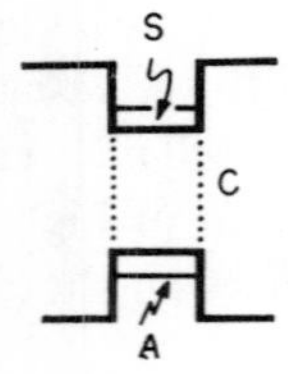

FIGURE 7–5 Junctional unit, see text. (Taken with kind permission from W.R. Loewenstein, 1966.)

69.000). Ca^{++} appears to have a double role: its total removal causes uncoupling of the electrical communication, but too much Ca^{++} increases the resistance of the intercellular path. In tissues like salivary gland and renal epithelia, the cells are interconnected through these junctions to all their nearest neighbors. In the urinary bladder and sensory epithelia, instead, communication appears to be more restricted, manifestating itself in small groups or chains of interconnected cells (Loewenstein, Socolar, Higashino, Kanno and Davidson, 1965). As a result of the cellular contacts described in this section ion move rather freely from cell to cell, but not from cell interior to exterior or across the whole epithelial membrane.

7.2 *Some effects due to composite membranes*

In order to cross an epithelial membrane through a transcellular route, a molecule should often traverse at least two cell membranes using in each instance some of the mechanisms we studied in chapters 3 to 6. Besides, each membrane can be crossed by the same substance through more than one mechanism arranged in parallel. Kedem and Katchalsky (1963a, b and c) have derived equations to describe the transport behaviour of membranes with more than one element arranged in parallel or in series. Among the conclusions they reach, one that directly concerns us here is that the behaviour of a composite membrane may deviate markedly from the one that could be expected from simple additivity of the properties of the individual elements, and only under very special circumstances the system can be described by linear equations of the type discussed in chapter 3 (see The Approach Based in Thermodynamics of the Irreversible Processes, page 67). An idea of the sort of effect that a multimembrane system can originate may be illustrated by an experiment carried out by Curran and McIntosh (1962). These workers have analyzed a system of three aqueous compartments in series separated by two artificial membranes where the central compartment may represent an epithelial cell. Curran and McIntosh demonstrated both, theoretically and experimentally that under certain conditions the system may transfer a net amount of water from the concentrated to the diluted solution. This mimics the behaviour of those epithelial membranes which take up water from the lumen and deliver it to the plasma.

For theoretical discussion of fluxes and forces and compartmental analysis of series membrane systems see Ussing (1958); Scheer (1960); Patlak, Goldstein and Hoffman (1963); Solomon (1964); Marro and Pesente (1964a and b); Parsons and Prichard (1966); Schwartz and Snell (1968).

7.3 Transepithelial transport

Our knowledge of epithelial membranes is intimately associated with the study of the frog skin through a series of historical circumstances. It started more than a century ago with Du Bois Raymond's observation that a frog skin can originate an electrical potential difference of more than 100 mV between the outer and the inner side. Galeotti (1904, 1907) demonstrated that this potential depends on the presence of Na^+ or Li^+ in the bathing solution. He correctly attributed the electrical behaviour to a higher Na^+ or Li^+ permeability in the inward than in the outward direction. However, on the assumption that it would violate the Second Law of Thermodynamics, this interpretation was—to put it mildly—disregarded. Thirty years later, Huf (1935) demonstrated that the frog skin can actually transport a net amount of electrolytes from one Ringer's solution to the other. Krogh (1937, 1938) later demonstrated that the frogs are able to take up sodium chloride from the surrounding medium even if the concentration in the latter is as diluted as 10^{-5}M. The discovery by Ussing (1949a and b) that the frog skin can take up a net amount of Na^+ from an outside Ringer's solution diluted a hundred times, definitively pointed the process of Na^+ translocation as one of the central process of the mechanisms to transport substances across epithelial membranes. The interest on the central role of Na^+ was enhanced by the rediscovery of Reid's observation (1902) that the presence of Na^+ either increases or is an absolute requirement for the active transport of aminoacids, sugars and other non-electrolytes (Riklis and Quastel, 1958; Csáky and Zollicoffer, 1960; Csáky, 1961; Schultz and Zalusky, 1963; Curran, 1965; Crane, 1965; Alvarado, 1966). Actually, most experimental models of the translocation of substances across epithelial membranes are directly or indirectly derived from a model proposed by Koefoed-Johnsen and Ussing (1958) to explain the transport of Na^+ across the frog skin.

It seems therefore convenient to focus our attention in the movement of Na^+ across the frog skin to obtain some insight of how epithelia work.

7.4 *Transcellular models for the transport of Na^+ across frog skin*

When mounted as a flat sheet between two Ringer's solutions, the frog skin develops an electrical potential difference between its inner (positive) and its outer side. Fukuda (1942) has shown that the presence of K^+ on the inside and Na^+ on the outside was a necessary condition to maintain this potential. This dependence was investigated in detail by Koefoed-Johnsen and Ussing (1958) in skins bathed with Ringer's solution in which Cl^- was replaced by an equivalent amount of $SO_4^=$ which is a non-permeable anion. By changing the concentration of Na^+ or K^+ on the outer or the inner bathing solutions these workers demonstrated that the potential changes by about 58 mV per a ten-fold change in the concentration of Na^+ on the outside or the K^+ concentration on the inside bathing solution. Changes of the K^+ concentration on the outside, or Na^+ on the inside do not appreciably modify the electrical potential. These behaviours of the outer and the inner side resemble quite closely that of reversible electrodes of Na^+ and K^+ respectively. This was taken to indicate that there is a compartment in the skin limited on the outside by a Na^+ sensitive barrier and on the inside by a K^+ sensitive one. Since Meyer and Bernfeld (1946) as well as Ussing (1948) suggested that the seat of the skin electrical potentials and salt-transporting mechanism was located in the cells of the *stratum germinativum* (the innermost layer of epithelial cells), the inner and the outer barriers were identified with the inner and outer facing membranes of these cells.

The way in which Na^+ transport participates in electrical phenomena was investigated by Ussing and Zerahn (1951) using the technique illustrated in figure 7–2. This figure depicts a frog skin mounted as a flat sheet between two chambers containing *identical* Ringer's solution on both sides. Two salt bridges connected to a high input impedance voltmeter measure the electrical potential. Two Ag–AgCl electrodes are used to pass current through the skin. The influx and the outflux of Na^+ can be simultaneously measured by using two different tracers of sodium (Na^{22} and Na^{24}) added to the outer and the inner chambers respectively. Ussing and Zerahn measured both: the amount of current that should be passed through the skin to cancel the electrical potential, and the net flux of sodium measured as a difference between the influx and the outflux. They have shown that under steady state conditions there is a larger inward than outward flux of Na^+—thus originating a net influx—, and that the charge moved per unit time by the net flux sodium is equal to the electric current necessary to abolish the electrical potential. Since the use of identical solutions on both sides, and the cancellation of the electrical potential with the short circuit current insure that no external driving force is present, the net inward movement of Na^+ is attributed to active transport. Zerahn (1956) and Leaf and Renshaw (1957) have shown that in order to maintain this current (*i.e.* the net transport of Na^+) an epithelial membrane consumes an extra amount of oxygen. The short circuit technique has since been extensively used to measure active transport across membranes. Although in the particular case of the frog skin, the urinary bladder (Leaf, Anderson and Page, 1958) and other membranes, the net flux of Na^+ is equivalent to the short circuit current, this is not necessarily the case with all other membranes. The magnitude of the short circuit current is given by the sum of the net transfer of *all* ions across the membrane, expressed as charge moved per unit time. It may happen that Cl^- is also transported in the same direction as Na^+ making the short circuit current lower than the net Na^+ flux (Zadunaisky, Candia and Chiarandini, 1963). In case Na^+ and Cl^- movement are equal and in the same direction, there may be no short circuit current at all (see for instance Diamond, 1962). Even in the case that Cl^- were not transported, only in the case of complete asymmetry (outer barrier impermeable to K^+, and inner barrier impermeable to Na^+) will short circuit current measure net Na movement (Ginzburg and Hogg, 1966).

7.5 The model of Koefoed-Johnsen and Ussing

On the basis of the observations mentioned above Koefoed-Johnsen and Ussing (1958) postulated the model shown in figure 7–6. It represents a cell of the *stratum germinativum.* The outer membrane of this cell was supposed to be permeable to Na^+, impermeable to K^+ and, when bathed in $SO_4^=$ Ringer's solutions, to behave as Na^+ electrode. The inner membrane was supposed to be permeable to K^+, impermeable to Na^+ and, when bathed in $SO_4^=$ Ringer's solutions behave as a K^+ electrode. This cell has a higher concentration of K^+ than the bathing solution due to a K–Na pump located at

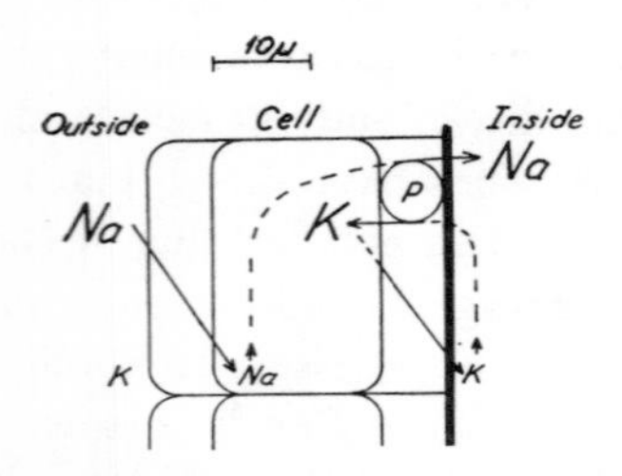

FIGURE 7–6 A stratum germinativum cell illustrating the hypothesis of Koefoed-Johnsen and H.H. Ussing, 1958.

the inner facing membrane. This exchange of K^+ by Na^+ keeps the cellular concentration of Na^+ low thus establishing a Na^+ concentration gradient across the outer barrier. This gradient provokes a net flux of Na^+ from the outside to the cell. The K^+ accumulated into the cell by the pumping mechanism can diffuse back toward the inside, but not toward the outside since only the inner facing membrane is permeable to K^+. Although under certain experimental conditions the pump can work electrogenically (see page 154), under physiological conditions the pumping is assumed to operate neutrally (*i.e.* it exchanges one Na^+ for one K^+) (Bricker, Biber and Ussing, 1963; Essig, Frazier and Leaf, 1963; Curran and Cereijido, 1965). The process of Na^+ transport across the frog skin depends, therefore, of two basic mechanisms: a passive diffusion across the outer barrier followed by an active extrusion at the inner barrier. Notice that the ion transporting mechanism that the cell uses to maintain its own ion balance is the same it uses to transport Na^+ across the epithelium. Now we are in the position to understand another important condition that the system should fulfill in order to measure net ion movements by means of the short circuit current: *it must be in steady state.* If, for instance, Na^+ enters from the outside and accumulates into the cell and the cell runs down of K^+ because the K^+ pumping does not match the leak of K^+ toward the inside, part of this leak will be included in the net current and erroneously attributed to Na^+ transport. As a transient, the short-circuit current can be measured even when the ion pumps are not functioning (Ginzburg and Hogg, 1966).

In the transcellular model, the electrical potential across the epithelial membrane bathed in $SO_4^=$ Ringer's solutions is thought to be the sum of the diffusion potential of Na^+ across the outer barrier and the diffusion potential of K^+ across the inner barrier. Na^+ does not contribute to the potential across the inner barrier because this barrier is not permeable to Na^+. For an analogous reason K^+ does not contribute to the potential at the outer barrier. The total potential, $\Delta\psi_{oi}$ is thus given by

$$\Delta\psi_{oi} = \Delta\psi_{oc} + \Delta\psi_{ci}$$

$$\Delta\psi_{oi} = \frac{RT}{F}\left[\ln\frac{(\mathrm{Na})_o}{(\mathrm{Na})_c} + \ln\frac{(\mathrm{K})_c}{(\mathrm{K})_i}\right]$$

where the subindexes *o*, *c* and *i* refer to the outer solution, the cell and the inner solution respectively (Koefoed–Johnsen and Ussing, 1958). In general agreement with this view, studies of the profile of the electrical potential across the epithelium was found to consist of two main steps. The first one is attributed to the Na^+ component and the second to the K^+ component (Engbaek and Hoshiko, 1957; Frazier, 1962; Cereijido and Curran, 1965; Whittembury, 1964).

7.6 *Revised version of transcellular models*

The net flux of Na^+ across the epithelium is a function of the concentration of Na^+ in the outer bathing solution (Kirschner, 1955; Morel, 1958; Cereijido, Herrera, Flanigan and Curran, 1964; Frazier and Leaf, 1964). As the concentration is rised the net flux goes up until it reaches a maximum value. Figure 7–7 illustrates an experiment carried out by Kidder, Cereijido and Curran (1964) which shows the increase in the electrical potential produced by a sudden increase in the Na^+ concentration of outer bathing solution from 2.4 to 47 mM. The new potential is reached in less than 600 milliseconds. Short circuit current would follow a similar time course. This delay is due to the time it takes for the Na^+ to diffuse and achieve the new concentration (47 mM) at the level of the Na^+ sensitive barrier.

By using a solution of the diffusion equation Kidder *et al.* measured the thickness of the layer interposed between the place where the new solution was put (*i.e.* the lucite outer chamber) and the Na^+ sensitive barrier. Since most of this thickness (some 20–25 μ) was accounted for by a non-stirred layer of outer bathing solution, they concluded that the Na-sensitive

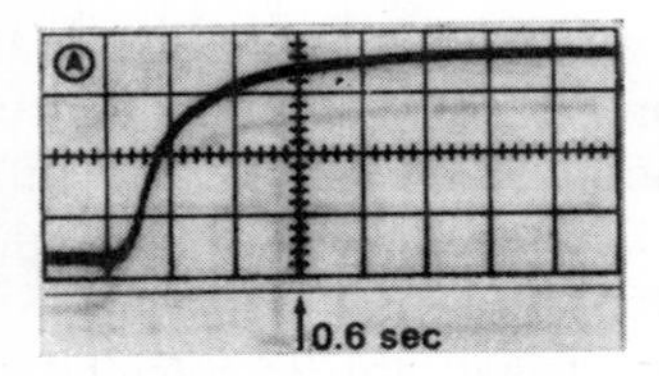

FIGURE 7–7 Reproduction of a original oscilloscope record showing the electrical response upon raising the Na concentration of the outside solution from 2.4 to 47 mM. One major divition on the vertical scale is 20 mV. On the horizontal scale is 200 msec. (From G.W. Kidder, M. Cereijido and P.R. Curran, 1964.)

barrier coincides with the outer anatomical border of the epithelium (see also Dainty and House, 1966). This information together with the observation of Curran and Cereijido (1965) that all the K^+ in the skin exchanges with K^{42} in the inside solution with apparently a single time constant, suggest that the outer barrier of the model of Koefoed–Johnsen and Ussing which was Na-sensitive and K-impermeable, must be located at the outer anatomical border of the epithelium. Therefore, present transcellular models proposed by Ussing and Windhager (1964) and Farquhar and Palade (1966) (Fig. 7–8) are not confined to the cells of the *stratum germinativum* but include also all the cells in the epithelium. According to these models Na^+ movement and distribution have three main characteristics: 1) Na^+ is homogeneously distributed in a unique cellular compartment owing to essentially free diffusion from cell to cell through the intercellular junctions of the type depicted in figure 7–5. 2) Na^+ is assumed to enter into the cell by diffusing down its gradient across the outer anatomical border of the epithelium. 3) The active step in Na^+ net movement across the skin is assumed to be located at the boundary between the cells and the interspace. Since this interspace is closed toward the outside but open toward the inside, the Na^+ pumped by the cells diffuses toward the inner bathing solution. Therefore, in this model the *outer barrier* is identified with the outer anatomical border, and the *inner barrier* with all the cellular borders facing the interspace. This view lead to more extensive reinvestigation of the profile of the electrical potential which was found to consist of 3 or more steps, each one thought to be given by a cell layer (Ussing and Windhager, 1964; Biber, Chez and Curran, 1966).

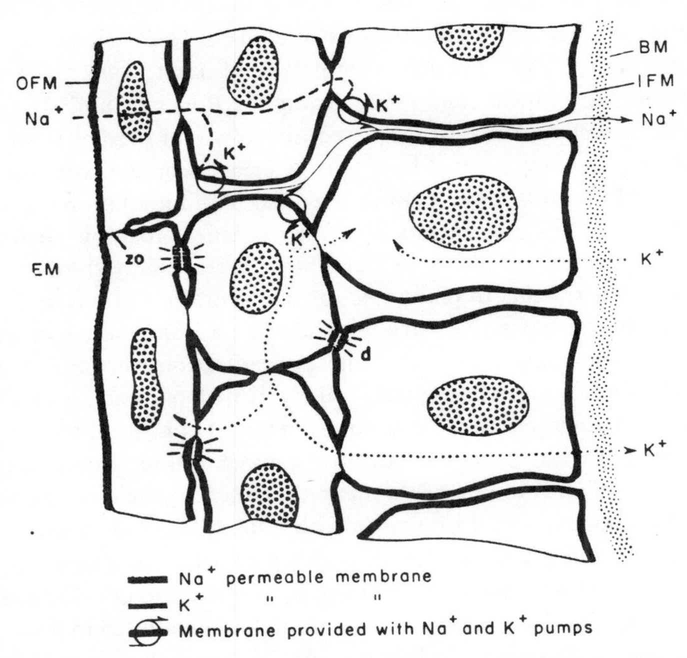

FIGURE 7–8 Transcellular model of Na^+ transport across frog skin. OFM: outer facing membrane. IFM: inner facing membrane. BM: basal membrane. ZO: zonulae occludens. d: desmosome. (Taken with kind permission from M. Farquhar and G. Palade, 1966.)

7.7 *Models with specialized transporting compartments*

As discussed above, the technique of mounting epithelial membranes with identical Ringer's solution on both sides and with the electrical potential difference abolished by the use of a short circuit current is an extremely useful method designed by Ussing and Zerahn mainly to cancel any difference in electrochemical potential between the outside and the inside, thus putting in evidence active fluxes. Transcellular models adequately describe Na^+ movement under these conditions. However, very seldom the outer facing membrane is in contact with a solution in which Na^+ is as concentrated as in a Ringer. In the particular case of the frog skin the outer surface is in contact with the water in the pond. Krogh has shown that the frog skin can take up NaCl from an outside bathing solution as diluted as 10^{-5}M. Modern studies have also shown that the skin, as well as other epithelia, can take up a net amount of Na^+ from solutions as diluted as 1 mM (Ussing, 1949a and b; Kirschner, 1955; Frazier, Dempsey and Leaf, 1962; Cereijido, Herrera, Flanigan and Curran, 1966; Rutunno, Pouchan and Cereijido, 1966). The Transcellular models discussed above would predict that, since Na^+ enters the cell compartmnet by a passive mechanism the concentration in the cell should be still lower than in the outer bathing solution (*i.e.* lower than 1 mM). Since this concentration seemed unreasonable low, Rotunno, Pouchan and Cereijido (1966) and Cereijido and Rotunno (1967) carried out a series of experiments to determine the concentration of Na in the epithelium when it transports Na^+ from an outside bathing solution with 1 to 10 mM Na. They have shown that this concentration can be as high as 97.0 mM. Computation of data published by Huf, Doss and Wills (1957) indicated that in *Rana pipiens* it can be as high as 114 mM. Since the magnitude of the electrical potential across the outer border cannot explain how Na^+ enters in a net amount into an epithelium 100-times more concentrated that the bathing solution, the only alternative seemed to assume an active step at the outer border. Active Na^+ transport in frog skin depends on a Na–K-ATPase sensitive to ouabain (Koefoed–Johnsen, 1957; Bonting and Caravaggio, 1963). Yet, as mentioned above and illustrated in figure 7–9, at the outer border there is no ATPase at all, Na–K sensitive or not. When the skin of the frog *Leptodactylus ocellatus* is bathed in isotonic Ringer's solution with 1 to 10 mM Na the net flux of Na is 0,50 μmole h^{-1} cm^{-2} (Rotunno, Pouchan and Cereijido, 1966). Since the skin transports 3 Na^+ per molecule of ATP split, it will take 0.167 μmole ATP h^{-1} cm^{-2} to account for the active flux. On the basis of the specific activity of the Na–K–ATPase of this tissue (0.182 μmole of P_i hydrolyzed per hour and per mg of protein) it was calculated that it takes more than two layers of cells to afford the necessary Na–K–ATPase to pump the 0.50 μmole of Na^+. Therefore, if—as required by the transcellular models discussed above—the epithelial Na were contained in a unique homogeneous compartment, the following unsolvable situation would arise: while an active step is required at the outer border, the necessary pumping mechanism is not there but scattered over more than half of the epithelium. Rotunno *et al.* suggested that the problem could be solved if epithelial Na were compartmentalized and only one fraction were involved in Na^+ transport across the epithelium. The existence of more than one Na-compartment has since been proved by a nuclear magnetic resonance analysis which demonstrated that a large fraction of Na is not free as Na-ion but bound (Rotunno, Kowalewski and Cereijido, 1967) and

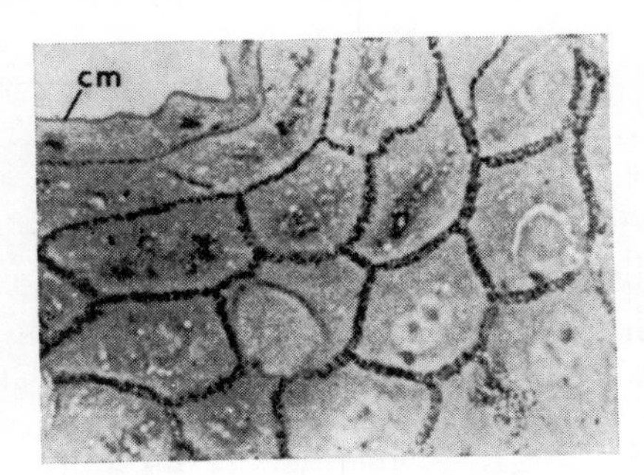

FIGURE 7–9 Photomicrograph (phase contrast) of an ATPase reaction in the frog skin. cm: cell membrane at the outer boıder. (Taken with kind permission from M. Farquhar and G. Palade, 1964.)

by a tracer kinetics analysis which indicated that *only a small fraction of the total Na* participates in the process of Na^+ transport (Cereijido and Rotunno, 1967).

In order to account for these findings and to afford a plausible explanation to the location of the active step discussed above, Cereijido and Rotunno (1968) introduced a new model for the transport and distribution of Na^+. We shall now summarize the main characteristics of this model, but before we start its description it would be convenient to mention some points that were taken into consideration in devising the new model.

1) As said above, epithelial Na is compartmentalized. Since only a small fraction of the total Na is involved in Na^+ transport *across* the epithelium (let us call it *Transporting Compartment*) then two main alternatives could solve the problem of how to transfer Na^+ *passively* from an outer solution with low concentration of Na^+, to an epithelium with a high overall concentration of Na: (a) that although the overall concentration of Na in the epithelium is high, the concentration in the transporting compartment were indeed lower than in the outer solution; (b) that the Na contained in the transporting compartment were not as a dissociated ion, but were bound to fixed charges.

2) The transport from the outer to the inner solution indicates that sooner or later Na^+ has to go through an active step. As discussed above, the Na–K-ATPase assumed to be the pump is contained in more than one layer of cells. Therefore the transporting compartment has to be distributed across of most of the epithelium.

TABLE 7-1 Unidirectional fluxes of sodium[a]

	pmole sec^{-1} cm^{-2}	References
Red blood cells (human)	0.11	Glynn (20)
Abdominal muscle (frog)	5	Keynes (28)
Sartorius muscle (frog)	10	Keynes (28)
Nerve (Sepia and Loligo)	33	Hodgkin and Keynes (23)
Urinary bladder (toad)	140	Frazier and Leaf (19)
Skin (frog)	834	Zadunaisky, Candia, and Chiarandini (55)
Skin	415	Curran and Gill (12)
Skin	390	Ussing (46)
Ileum (rabbit)	2760	Schultz and Zalusky (42)
Ileum (rat)	1330	Curran (9)
Cecum (guinea pig)	2160	Ussing and Andersen (49)

[a] Taken in the active direction. *Outfluxes:* red blood cells, muscles and nerve. *Influxes (or mucosal to serosal side):* urinary bladder, skin, ileum, and cecum. [Taken from Cereijido and Rotunno (1968)]

3) Table 7–1 shows the value of fluxes in several biological systems. In single cells the flux takes place across one cell membrane only. In epithelial membranes, —according to transcellular models—it crosses two cell membranes: one to get in and another to get out at the opposite side of the epithelium. Notice that Na^+ crosses epithelial membranes more easily than single plasma membranes. Although this might be just fortuitous, it seems odd that Na^+, whose ability to cross lipid bilayers is so poor (see page 145) goes so easily across *two* cell membranes. In designing the new model a route was sought in which Na^+ would passively enter into an epithelium with high Na content, would not have to cross necessarily two lipid bilayers, and could reach the active step located deep into the epithelium.

We shall now describe the generel characteristics of the new model. Figure 7–10 depicts two cells of a frog skin in which the lipid components of the cell membrane were blown up to explain the basic idea of the new model. The model assumes that the Na-selectivity of the outer facing membrane is due to a particular ionic strength of the polar groups of the outer leaflet of the cell membrane. The ability of any ion to penetrate lipid bilayers is so poor that for a Na ion coming from the outer bathing solution it will be easier to travel tangentially by jumping from fixed polar group to fixed polar group, than to cross the hydrophobic component of the membrane; *i.e.* it is easier for the ion to go around rather than penetrate the cell. It passes through the *zonulae occludens* and reaches the Na-selective polar groups in the inner facing membrane. It crosses to Na-selective polar groups of the cell membrane of deeper layers of cells by travelling over the fixed charged groups of the pillars which compose the extracellular material of the desmosomes (assuming that they have a structure similar to the one depicted in figure 7–3). The model assumes that the barrier of very low Na-permeability of the inner facing membrane, proposed first by Koefoed–Johnsen and Ussing and supported later by experiments of MacRobbie and Ussing, (1961) and by studies of the intracellular electrical potential, is located between the polar groups where Na^+ migrates and the intercellular

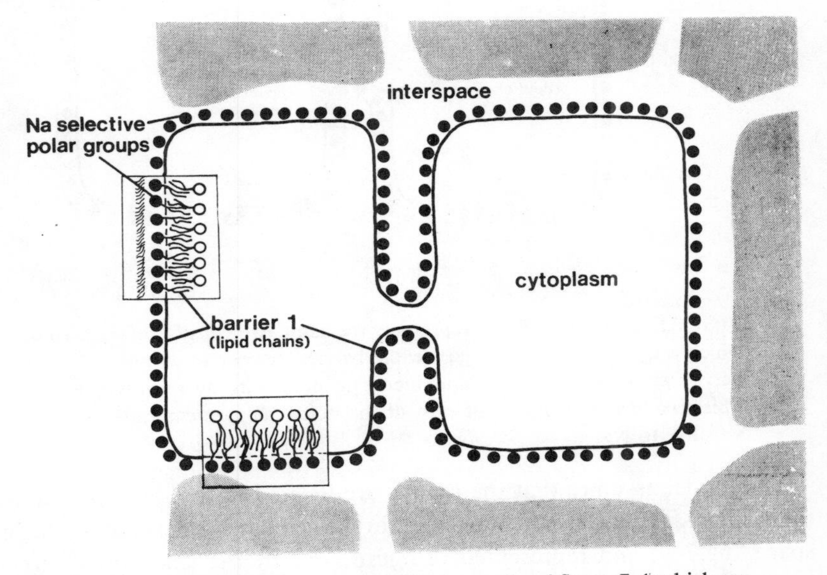

FIGURE 7–10 This figure represents the two cells of figure 7–1 which are not shadowed. The lipid component of the cell membrane is blown up and the polar groups of the outer leaflet are represented as full circles. The bold line represents the hydrophobic component of the cell membrane (barrier 1).

space (fig.7–11). This would mean that the Na-selective polar groups are sandwiched between two barriers of low Na-permeability: the hydrophobic chains of the lipids (Barrier 1), and the barrier which confers to the inner facing membrane its low Na-permeability (Barrier 2). Since the outward facing membrane is sensitive to Na^+, the model assumes that barrier 2 is lacking at the outward facing membrane where the groups can be easily reached by Na^+ ions. The bold lines represent the Na-impermeable barriers. Only at the outer anatomical border could Na^+ freely enter or leave the compartment limited by the two barriers.

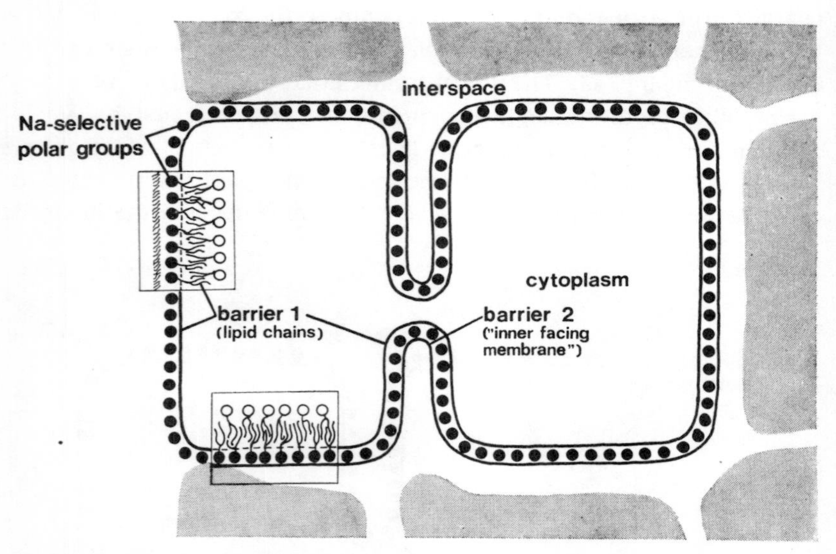

FIGURE 7–11 This figure represents the same two cells of figure 7–10 in which a hypothetical Na-impermeable barrier (barrier 2) was added. This barrier would confere to the inner facing membrane its low sensitivity to Na. This second barrier does not exist at the outer facing membrane which is sensitive to Na.

The model assumes that the pumps are located in the same position that the ATPase is found hystochemically: in the cellular borders facing the interspace (fig.7–9) and represented in figure 7–12 by the small dots. Now the basic idea of the model is completed. Let us now illustrate how it works.

The pumps located over the inner facing membranes translocate Na^+ from the Na-selective polar groups to the intercellular space across the Na-impermeable barrier. Since the empty site left by a pumped Na^+ cannot be easily reached by Na^+ from the interspace of from the cells, the site is refilled when the Na^+ acting as a counterion of a neighbor polar group jumps to the empty site. The overall effect is to move the empty site sidewise towards the outward facing membrane where another sodium can be adsorbed (see lower part of figure 7–12). Notice that this mechanism is essentially independent of the one used by the cell to maintain its sodium balance, a suggestion supported for instance, by the demonstration of Huf, Doss and

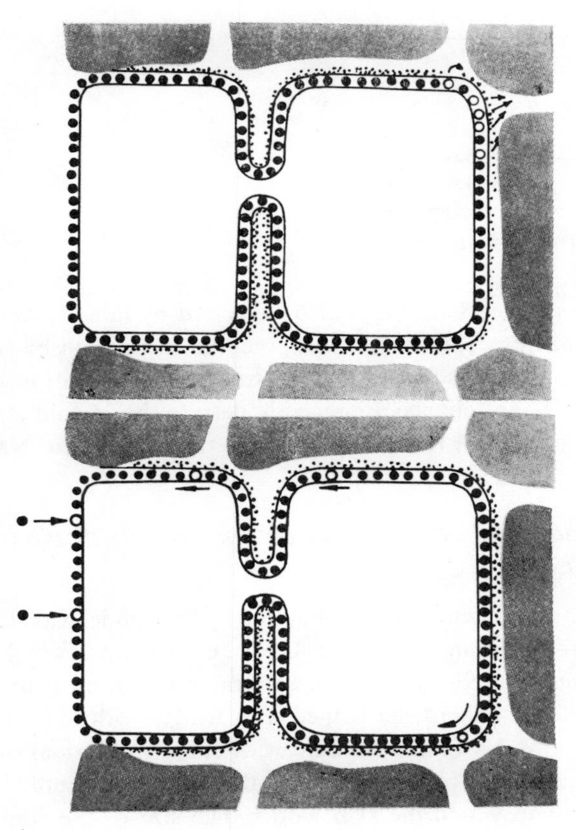

FIGURE 7–12 This figure represents the same two epithelial cells of figure 7–10 and 7–11 in which the Na pumps were added in the same position in which the ATPase is found, *ie*. the cell borders facing the interspace (see figure 7–9). The pumps translocate Na toward the intercellular space from where it diffuses toward the inner solution. The empty site left by a pumped Na (empty circles) is refilled by sidewise migration. See text.

Will (1957) and Levinsky and Sawyer (1953) that metabolic inhibitors or diuretic substances can stop the net transport across without disturbing the electrolyte balance of the cell (see also Curran and Cereijido, 1965). The

TABLE 7–2 Effect of sodium concentration on cell potential under short-circuit conditions

Na concentration	PD (cell-outside)	Short-circuit current
mM	mV	μA/cm^2
115	-16.0 ± 4.8	58
15	-16.7 ± 5.1	36
8	-16.5 ± 5.0	26
3	-16.9 ± 4.6	12

From Cereijido and Curran (1965)]

independence of the two mechanisms is also suggested by table 7–2. It belongs to a study of the intracellular electrical potential carried out by Cereijido and Curran (1965) in which they demonstrated that the intracellular electrical potential does not change appreciably despite the 36-fold change in the concentration of Na^+ or the 4-fold change in the transport of Na^+.

7.8 *Size and connections of the compartments of Na in the epithelium*

The detailed compartmentalization of Na required by this model was studied by Cereijido, Reisin and Rotunno (1968) in slices of epithelium in which they analyzed the distribution of Na by chemical and tracer kinetics. Figure 7–13 summarizes their findings in epithelia bathed with Ringer with 115 mM Na on the inside and 1 mM Na (made isotonic with choline chloride) on the outside. Among the different conditions tested this resembles more closely the natural conditions in which the skin works. The size of the compartments in the drawing is proportional to the *amount* of Na they contain. However, the size does not represent the *volume* occupied by the compartment in the epithelium. Thin lines represent boundaries which have a comparatively high Na permeability. Only 8 to 13% (0.020 μmole mg^{-1} of dry weight) of the total Na in the epithelium is contained in the transporting compartment (T). This compartment is bound on the inside by a Na-impermeable barrier. About one third (0.070 μmole mg^{-1} of dry weight) of the total Na is contained in a compartment which does not appreciably exchange its Na with Na^{22} in the solution in 40 to 80 minutes. Since fluxes across equilibrate in much less than 40 minutes this compartment does not seem to be involved in the movement of Na^{+} across the epithelium. The rest

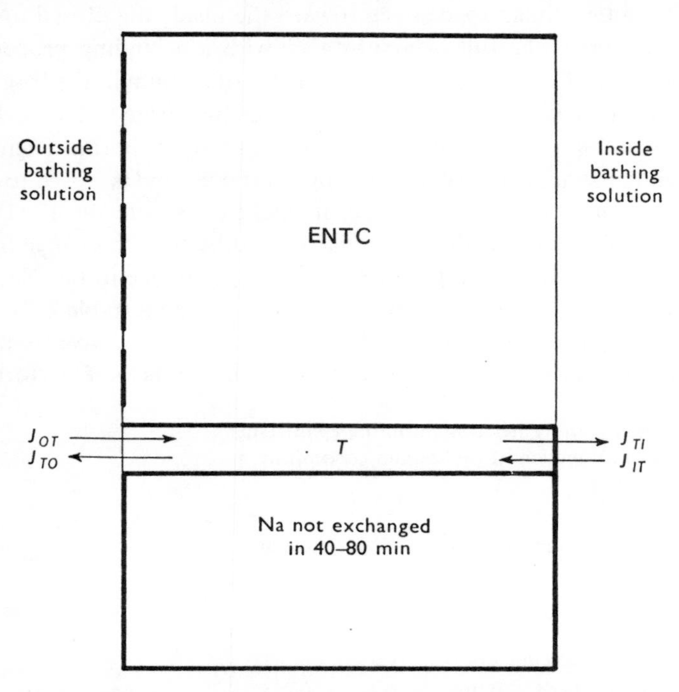

FIGURE 7–13 Model used to analize the movement and distribution of Na in the epithelium of the frog skin. T is the sodium pool in the transporting compartment. The size of the compartments in the drawing is proportional to the *amount* of sodium they contain. However, the size does not represent the *volume* occupied by the compartment in the epithelium. (From M. Cereijido, I. Reisin and C. A. Rotunno, 1968.)

of the epithelial Na (0.134 μmole mg^{-1} of dry weight) is contained in a compartment called exchangeable-non-transporting-compartment (ENTC) which exchanges its Na with the inside solution and which does not appear to be directly involved in Na transport *across* the skin.

We may now compare figures 7–13 with 7–1 and 7–12 and discuss briefly where the compartments might be located. MacRobbie and Ussing (1961) have clearly shown that changes in Na concentration on the inside solution fail to produce variations in the volume of the epithelium. It was also mentioned above, that those changes do not modify the electrical potential across the inner facing membrane. However the compartmental analysis discussed above indicates that the Na content of the epithelium is drastically changed. The intercellular space open toward the inside but closed toward the outside appears to be full of a substance with ion binding properties. Besides, in a study of nuclear magnetic resonance of sodium in the frog skin Rotunno, Kowalewski and Cereijido (1967) demonstrated that a large fraction of the sodium in the skin is bound and not free as Na-ion. Therefore if the intercellular space is full of ion binding substance, and is closed toward the outside but open toward the inside, it might constitute the ENTC of figure 7–13. In connection with this, notice that although the change in the concentration of Na of the Ringer produces a large change in the Na content, the amount of water in the epithelium remains constant (table 7–3). Part of this space might also be accounted for by part of connective tissue included in the slices of epithelium where the measurements were performed.

TABLE 7–3 Sodium and water content in the epithelium of the frog skin as a function of the sodium concentration in the bathing solution

Concentration of sodium in the bathing solution	Sodium content	Water content
(mM)	(μmole/mg dry wt)	(μl/mg wet wt)
115	0.264 ± 0.010	0.763 ± 0.005
1	0.084 ± 0.010	0.755 ± 0.005

[From Cereijido, Reisin and Rotunno (1968)]

In discussing the Na not-exchanged in 40–80 minutes we will again take into consideration the very low Na^+ permeability of the inner facing membrane. If the epithelial cells depend on the inner facing membrane to maintain their balance of Na, it would be expected that the cellular Na constituted a very slow compartment. Therefore the 0.070 μmole of Na per milligram of dry weight which does not exchange in 40 to 80 minutes might constitute the intracellular Na. If this were so, and we assume that all the epithelial water (3–34 μl per mg of dry weight) is contained in the cells (*i.e.* we neglect that part of it, is in the narrow intercellular space), then the concentration of sodium in the cells would be 21 mM.

As with respect to the Na transporting compartment, the fact that it contains only 8–13% of the total Na in the epithelium, and is permeable to Na on the outside but impermeable on the inside agrees with the small transporting compartment confined to the cell membrane of the epithelial cells. The fact that the transporting compartment of this model is constituted by the cell membranes, should not be taken to implicate that it is confined to the *periphery* of the cells. Should the concept that the different membranes of a cell constitute a sort of continuum (see page 190) be correct, it would be possible that the membranes forming infoldings, the membranes of the Golgi apparatus, the membranes of the nucleus, etc. would contain part of the transporting compartment.

7.9 Comments on the non-transcellular model

This model considers that epithelial cells have mechanisms for regulating their K- and Na-concentration levels as any other cell in the body. In this respect they are relatively independent of the hazards of the environment and enjoy the steadiness of the Internal Environment. The net transport across the epithelium may stop without disturbing the electrolyte balance of the cells. Their participation in the process of Na^+ transport arises as a specialization that requires a further degree of organization and is, therefore, more vulnerable. Thus, in any situation the conservation of the steady state of the cell would prevail. If the supply of energy is made scant by dinitrophenol, fluoracetate, iodoacetate, azide or diethylmalonate, the cells will lose their ability to transport Na^+ *across* the epithelium but will maintain their ion composition until the metabolic inhibition becomes more intense (Huf, Doss and Will, 1957; Curran and Cereijido, 1965). The independence of the two functions can be further illustrated by considering figure 7–14. In this study Curran and Cereijido (1965) measured simultaneously the flux of K^+ from the inside bathing solution to the cells, and the flux of Na^+ across the epithelium under two different conditions: with high (black dots) and with low (triangles) concentration of Na^+ on the outer bathing solution. The lines connect the two measurements performed on the same piece of skin. We observe first that the data do not fall in a straight line as it would be expected if Na and K transport were coupled as illustrated in figures 7–6 and 7–8. We also observe that when the Na^+ transport across decreases as a consequence of using a low Na concentration on the outside, the transport of K^+ is very little affected. This would indicate K^+ and Na^+ transports are *not* coupled. If, on the other hand, epithelial cells have a Na–K pump to regulate its Na^+ and K^+ content as most other cells do, then not only the transport of K^+, but also the transport of Na^+ out of the cell would be independent of the transport of Na^+ across as suggested by the non-transcellular model.

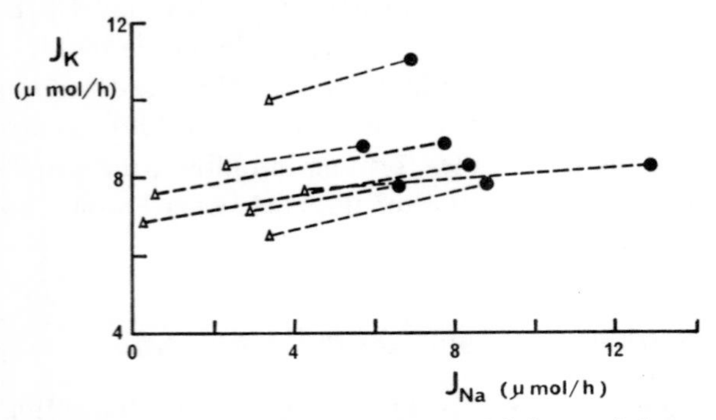

FIGURE 7–14 Relationship between K influx and the net flux of Na in frog skin. At high (dots) and low (triangles) concentration of Na in the outer solution. Dashed lines connect the measurements corresponding to the same membrane. (After M. Cereijido and P. F. Curran, 1965.)

The fact that the route of Na^+ transport of the model goes across the epithelium without entering the cytoplasm should not be taken to imply that the outward facing membrane of the epithelial cells is impermeable to Na^+. When the concentration in the outer solution is high enough or the electrical potential inside the cell is made negative with respect to the bathing solutions by means of a short circuit current (Hoshiko, 1961; Whittembury, 1964; Cereijido and Curran, 1965) the entry of Na^+ into the cell and its subsequent extrusion toward the interspace may play a significant role in the over-all Na^+ transport. It is conceivable that a substance capable of increasing the permeability of the outward facing membrane of the cells will put the transcellular path to work thus enhancing the net transport. Besides, as discussed below, the transcellular route may be involved in the coupling of Na^+ movement to the movement of other substances.

The explanation given here concentrates on how Na moves without describing what is the fate of the mobile anion. Notice that if the mobile anion were totally excluded, the epithelium will only have a *net* Na^+ flux when a short circuit current is passed (see chapter 6). However this is not the case most epithelial membranes have a spontaneous net (active) flux of Na which gives rise to a net (passive) flux of Cl^-, and some others have also an active transport of Cl^-. There is though, little information on the intrinsic mechanism used in Cl^- migration.

7.10 *The coupling of Na transport to the fluxes of other solutes*

As mentioned before, the accumulation of sugars and aminoacids within epithelial cells against a concentration gradient is coupled—at least partially—to the active transport of Na^+ across the epithelium. Studies of alanine transport in isolated rabbit ileum carried out by Curran, Schultz, Chez and Fuisz (1967) (see also Schultz, Curran, Chez and Fuisz, 1967) have indicated that, although the alanine flux is transported as a consequence of its coupling

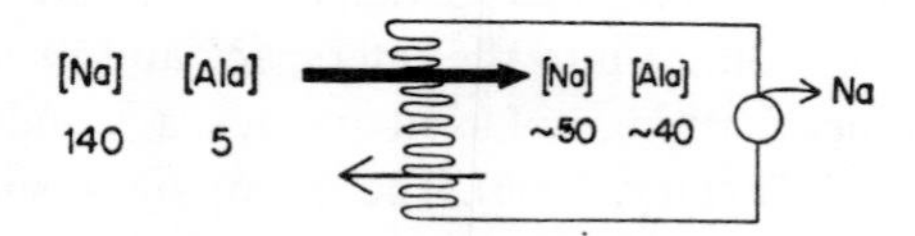

FIGURE 7–15 Alanine and sodium fluxes across intestinal mucosa. The numbers indicate the concentration of sodium and alanine. (Taken with kind permission from A.R.Chez, R.R.Palmer, S.G.Schultz and P.F.Curran, 1967).

to the *process* of active transport of Na^+, it is not coupled to the active *step* of this process. These workers have demonstrated that a large fraction of the alanine flux is coupled to the flux outside→cell (see figure 7–15). Na entry across this barrier seems to be a passive diffusion (from 140 to 50 mM) mediated by a carrier which translocates Na^+ and/or alanine, but which works faster if both of them are present. In this way the net influx of Na^+ across the outer border results in a net influx of alanine. Notice that the net flux at the outer border is a passive mechanism which operates as long as the Na pump at the other side of the cell manages to keep the cellular concentration of Na low. Hence alanine accumulates from 5 to 40 mM due to its coupling to the passive step of an over-all active process. Although to the best of our knowledge, truly unidirectional fluxes of both Na^+ and the coupled non-electrolyte in epithelia were only measured by Curran *et al.* using a special technique, important studies on the coupling, and suggestions on why it comes about were done by many other workers (see for instance Riggs, Walter and Christensen, 1958; Crane, 1962; Schultz and Zalusky, 1963; Alvarado, 1966). For an interesting study of this coupling in individual cells see Vidaver (1964).

7.11 The coupling of water and solute transport

Water transport is a basically passive mechanism osmotically coupled to the movement or distribution of solute. Epithelial cells take up solutes from the outside and by one route or the other pump it into the interspace. Under physiological conditions in most epithelia this solute is mainly NaCl, either because Na^+ and Cl^- come as primarily transported ions, or because the electroneutrality requirement in non-short-circuited preparations makes Cl^- to accompany Na^+. If there is only a small transport this may be carried out mostly by the outermost cell layer. As the amount of solute to be pumped increases (say because its concentration in the outer bathing solution is increased) other layers of cells would also participate (see figures 7–8 and 7–12). The continuous input of solutes into the interspace and its subsequent diffusion toward the inner bathing solution creates a standing-gradient flow system (Diamond and Tormey, 1966). The pumping of solute (white arrows) develops a high solute concentration which results in its diffusion toward the open mouth of the interspace and also in a high tonicity of the solution filling the interspace which drives water from the

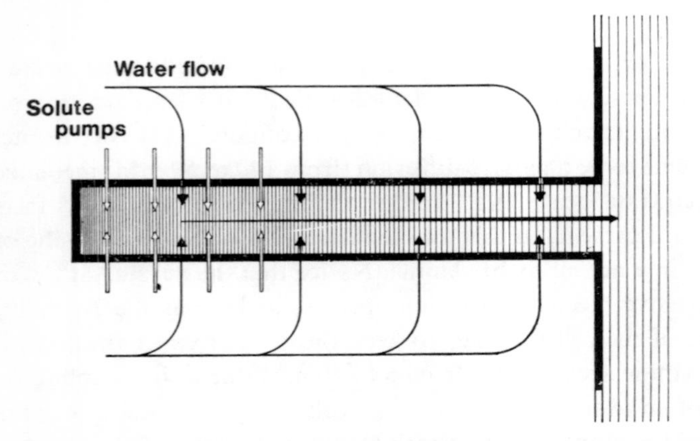

FIGURE 7–16 Diagram of a standing gradient flow system consisting of a long and narrow channel closed at one end. See text. (Taken with kind permission from J. Diamond and W.H. Bossert, 1967.)

surrounding. Diamond and Bossert (1967) have analyzed mathematically the behaviour of this system. The purpose of this analysis was to find the osmolarity of the fluid emerging in the steady state from the open end of the channel. They found that the osmolarity depends upon five parameters: channel length, radius, water permeability, solute transport rate, and diffusion coefficient of the solvent through the lumen of the channel. They assigned to these parameters values which are typical of epithelial membranes and succeeded in giving a plausible explanation of the water-to-solute coupling. In figures 7–17 and 7–18 only one parameter is varied at time, the others are kept constant. It can be observed that, as the length of the channel increases, the osmolarity of the emergent fluid decreases. This can also be intuitively understood in figure 7–16: the depth of the shade indicates the solute concentration, and it is obvious that if the length of the channel were shortened, the emerging fluid would be more concentrated. Figure 7–18 shows that as the length of the solute input increases (more cell layers recruited) the concentration of the emerging fluid increases. Although a detailed account of the analysis carried out by Diamond and Bossert will not be given here, their results may be summarized as follows. "Two forces are responsible for carrying solute out the open mouth of the system: the diffusion of solute down its concentration gradient, and the sweeping effect of water flow upon the solute. Any factor that increases the relative importance of diffusion will make the emergent fluid more hypertonic, while factors increasing the relative importance of water flow are associated with a more nearly isotonic emergent fluid." It should be noted that this mechanism permits the transport of hypertonic or isotonic fluid, never hypotonic. This agrees with the observation that in biological prepations the primary transported fluid may be either isotonic or hypertonic to plasma. Hypotonic secretions in vertebrates appear to result from isotonic secretion followed by hypertonic reabsorption.

We have now a basic idea of how epithelial membranes translocate solutes and water based mainly in the description of phenomena observed in frog skin, urinary and gall bladder, and intestine. For other examples of complex membranes see: *Thyroid* (Wolff, 1964), *Gastric mucosa* (Rehm, 1964; Durbin and Moody, 1965; Hogben, 1968; Villegas and Sananes, 1968) *Renal tube* (Giebisch, 1968; Whittembury, 1968), *Cornea* (Zadunaisky, 1966; Green, 1966; Candia, 1968).

For recent information on the functional role of the brush border of epithelial cells see Crane (1966) and Passow (1967).

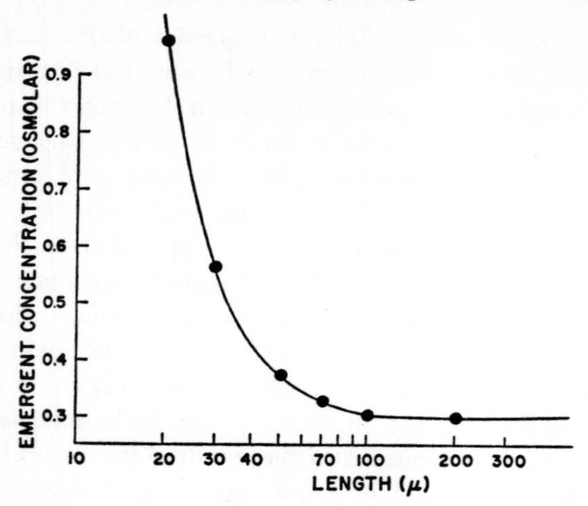

FIGURE 7–17 The effect of varying channel length upon the osmolarity of fluid produced by standing gradient flow system. (Taken with kind permission from J. M. Diamond and W. H. Bossert, 1967.)

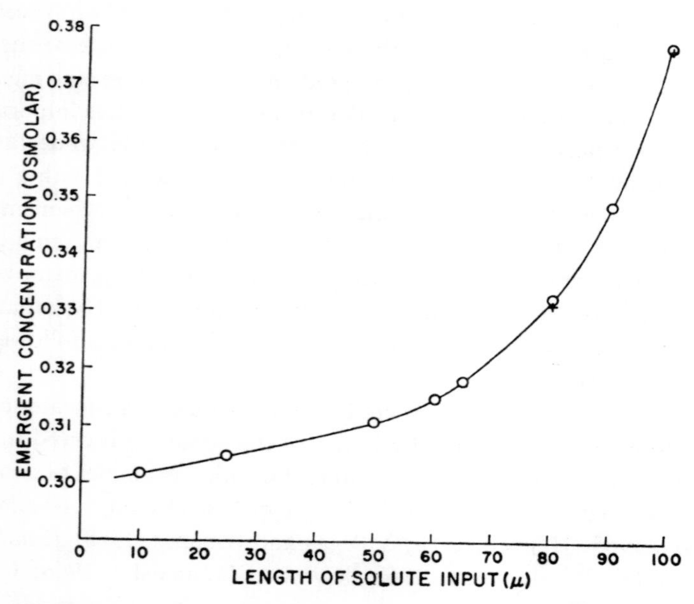

FIGURE 7–18 The effect of the length solute input upon the osmolarity of fluid produced by a standing-gradient flow system. (Taken with kind permission from J. M. Diamond and W. H. Bossert, 1967.)

8

Modification of the properties of biological membranes

ALTHOUGH IN chapter 1 we have learned that membranes could adopt several conformations and, eventually, switch from one to the other, in the rest of the chapters we have generally considered phenomena in which the properties of the membranes remain constant. In this chapter we shall consider some situations in which the membrane changes its behaviour or even its composition as a consequence of chemical changes in the environment, the presence of hormones, electrical stimulations, etc. We will successively describe inducible bacterial permeases, pinocytosis, the nerve impulse, and the effect of some hormones. This chapter is not meant to review these topics but to use them as illustrations of the ability of membranes to adequate themselves to functional requirements of a cell or a whole multicellular organism.

8.1 Permeases

As a general rule hydrophilic molecules diffuse very slowly through bacterial membranes. However, bacterial membranes, as any other membrane, select certain polar molecules which are essential for their survival. To explain this property carriers and pumps of the type described in chapters 4 and 5 were proposed. Figure 4–1 shows the hydrolysis of ortho-nitro-phenyl-β-D-galactoside (ONPG) by two groups of Escherichia coli (Cohen and Monod, 1957). Since in both groups of bacteria the rate of hydrolysis depends on the rate of penetration of (ONPG), these authors followed the penetration by measuring the rate of hydrolysis. In the cryptic bacteria, penetration is a linear function of concentration. In the normal group, penetration shows the characteristics of a carrier mediated system (Cohen and Monod, 1957; Hoffee, Enclesberg and Lamy, 1964; De Busk and De Busk, 1965; Koch, 1967). The fact that the operation of these systems leads in some cases to the accumulation of substances in the cell suggested the possibility that the system involved might be a pump (Kepes and Cohen, 1962; Koch, 1967). All these mechanisms for the non-simple-diffusion of substances into bacteria are usually called *permeases* (see Rickenberg, Cohen, Buttin, and Monod, 1956).

Should all these properties be the only properties that permeases had, they would had been included in chapter 4 and 5 among all others carriers and pumps. However some permeases have the peculiar characteristic of being inducible, *i.e.* if a new substrate is put in the culture medium, after a while one may observe that the bacterial membrane possesses a highly specific permease that was not there before (or immediately after) the addition of this substrate. That means that the information that a new substrate is available, was processed by the protein-synthesizing machinery, and the membrane becomes equipped with a new system so that the bacteria can take up the substrate. Theleologically speaking, it looks like Nature had the following reasons to devise inducible permeases: 1) the bacterial cell will not build an elaborated carrier or pump system, generation through generation, just in case a given substrate becomes available, but 2) in case the substrate happens to appear, and to constitute the only food-stuff, the bacteria should be able to take it efficiently enough as to growth. 3) Each permease system can translocate a particular substrate at a rate sufficient to support growth if it were the only substrate available (Kepes, 1963).

Efforts to extract a permease from the membrane were made by Fox and Kennedy (1965) and Kolber and Stein (1966). Figure 8–1 illustrates the basic idea of Fox and Kennedy's studies. N-ethylmaleimide (NEM) can bind and irreversibly inhibit the β-galactoside permease. Thiodigalactoside (as other substrates of the permease do) protects the active center of the permease system from the inhibitory binding of the NEM. When the cells are first allowed to react with NEM in the presence of thiodigalactoside NEM binds all around the cell membrane *but* on the active center of the permease. If then the galactoside is removed, and the cells are treated with NEM* (labelled), the active centers will bind the NEM* (labelled). Unfortunately, this method does not entirely prevent the labelling of some unspecific membrane components. To compensate this fact, Fox and Kennedy used this technique on two groups of cells: one group (cells with permease) was labelled with NEM–C^{14} and the other group (cells without permease) was labelled with NEM–H^3. They found a fraction of cell membrane protein in which the ratio C^{14}/H^3 (*i.e.* specific/non-specific labelling) had the highest value and which was assumed to contain the permease.

As said above, the isolation of the β-galactoside permease was also intended by Kolber and Stein. Oddly enough, this group ended up with a different material than the one extracted by Fox and Kennedy. It was suggested that the overall operation of the permease system involves more than one protein. For an excellent discussion of this problem see Stein (1967).

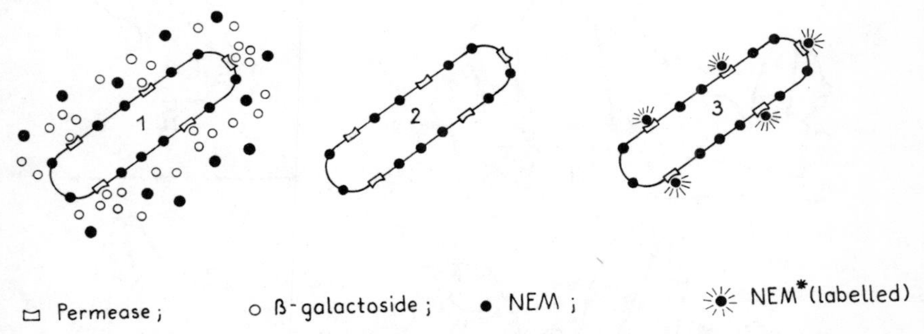

FIGURE 8–1 Three successive stages in the labelling of the β-galactoside permease of the *Escherichia coli*. 1) β-galactoside prevents the binding of NEM to the permease. 2) NEM is unspecifically attached to the membrane but not on the permease. 3) NEM* (labelled) attaches to the only sites left available by the non-labelled NEM: the permease.

8.2 *Pinocytosis*

Pinocytosis (cell drinking) was first described by Lewis (1931) and consists in the formation of small droplets of extracellular liquid surrounded by cell membrane (pinosomes) and which are pinched off from the surface. Figure 8–2 shows a drawing by Mast and Doyle (1934) in which several stages in the formation of the pinosomes can be seen from the emission of pseudopods (top left, *A*), until the pinosomes are released into the cytoplasm (bottom right, *D*). Fagocytosis, Pinocytosis and micropinocytosis (Palade, 1953; Palay and Karlin, 1959) are believed to be basically the same phenomena (Rustad, 1964). Novikoff (1961) proposes to call them "cytosis" without elucidation of whether the cell is eating or drinking. This process permits the penetration of solution which may contain large molecules which otherwise would be unable to diffuse through the membrane. Pinocytosis is traditionally associated to the penetration of proteins and other macromolecules. Thus Brachet (1957) observed intracellular RNA hydrolysis after the addition of RNAase to the bathing medium. Large molecules not only benefit from this type of permeation, but they also induce pinocytosis.

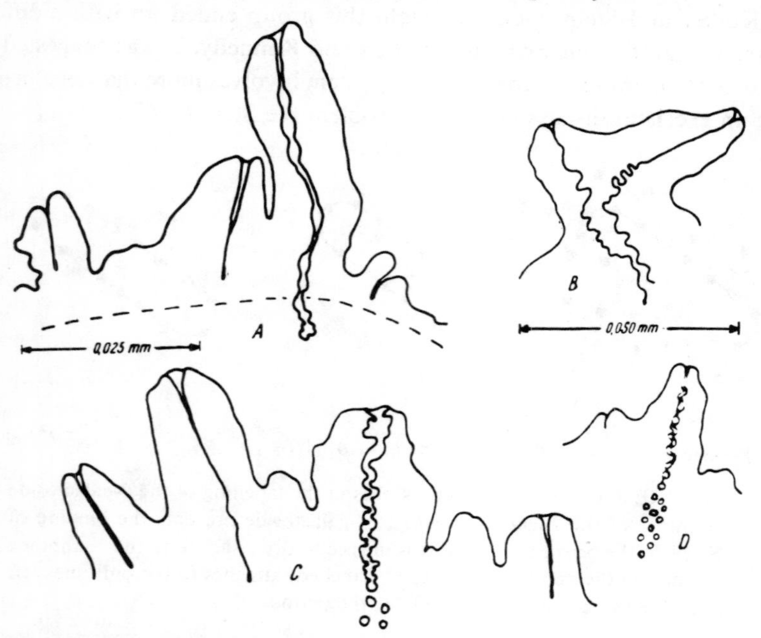

FIGURE 8–2 Schematic representation of different stages of pinocytosis. (After S.O. Mast and W.L. Doyle, 1934.)

Pinocytosis starts by the binding of large amounts of macromolecules to the outer surface of the cell (Brandt, 1958; Mercer, 1959; Schneider and Wohlfarth-Bottermann, 1959). The kinetic of this process as well as some competition observed between proteins and dyes, led to the comparison of the first stage of pinocytosis with an enzyme-substrate process (Odor, 1956; Gosselin, 1956; Brandt and Pappas, 1960). The induction of pinocytosis causes a drop in the resting potential of the amoeba (Josefsson, 1966; Brandt and Freeman, 1967). According to Chapman–Andresen and Holter (1955). In the presence of pinocytosis the amoeba becomes also permeable to glucose. It is not clear why adsorption results in the sinking and subsequent formation of pinocytotic tube, nor why the pinosome is detached from the tube. According to Shea and Karnovsky (1966), Brownian motion is sufficient to account for the migration of the vesicles in the greater part of their journey across the cell. Studies of time lapse cinematography carried out by Gey, Shapras and Borysko (1954), indicated that a moving pinosome can break a mitochondrion in its path (!).

The membrane forming the pinosome also becomes very much permeable once it is taken into the cytoplasm (Marshall, 1966). Small secondary vesicles form at expenses of the initial one. Primary vesicles can engulf smaller ones or form complex polyvesicular bodies. All the while substances inside the pinosomes seem to undergo digestion and products are subsequently released in the cytoplasm.

Pinocytosis is a function widely distributed. It was observed in plants, protozoa, higher animals, etc. (Bennett, 1956; Jensen and Mc Laren, 1960; Buvat, 1958; Rustad, 1964). It is particularly well developed in certain cells. An amoeba, for instance, can take in five minutes the RNAase contained in a volume of solution fifty times the volume of the amoeba. (Shumaker, 1958). This, by the way, gives an idea of membrane dynamics: if all this solution enters through pinocytosis, the cell will have to regenerate the membrane quickly enough so that it matches the amount of membrane removed with each pinosome.

According to Marshall (1966) the intracellular pool into which membrane disappears, and from which new membrane forms, compromises nine to ten times as much material as it is to be found in the surface membrane itself at any time.

Pinocytosis is inhibited by all those factors which interfere with metabolism: cyanide, carbon monoxide, low temperature (De Terra and Rustad, 1959; Chapman–Andresen, 1962; Karnovsky, 1962). In this sense it would constitute a case of active transport. But if this is so, then almost any substance could be actively transported, because the pinosome, besides of carrying the substance which promoted the pinocytosis, may also carry other substances present in the solution. Sbarra, Maney and Shirley (1961) (see also Sbarra, Shirley and Bardawill, 1962) have found that substances such as sodium malonate and γ-globulin added to the solution can penetrate only if the cells are phagocyting particulate material. They called this process "piggy back phagocytosis".

Some authors consider that the membranes participating in pinocytosis, phagocytosis, micropinocytosis, and secretion, as well as those forming the different cellular organelles, may be associated or represent different stages of a process by which membranes are synthesized and disassembled. For discussion of these points see: Palade (1956), Wellings and Deome (1961), Robertson (1962).

Robertson (1962) has suggested that the different membranes of the cell may represent the present condition of an evolutionary process by which the surface membrane enlarged, infolded and sorrounded (and gave rise to) the different organelles. In some cells, like for instance skeletal muscle cells, the surface membrane does constitute a continuum with some intracellular membranes.

Biophysicists assign an important role to the membranous tubular systems connected to or continuous with the peripheral cell membrane. In muscle cells for instance, they are thought to participate in the coupling between membrane phenomena and the mechanochemical events resulting in muscle contraction (see for instance Adrian and Freygang, 1962a and b; Freygang, Goldstein and Hellam, 1964 and Freygang, 1965).

8.3 Oscillatory phenomena in membranes—the nerve impulse

Although the operation of a carrier or a pump are cyclic processes, the flux across a membrane through such mechanisms is steady and time independent. However there are circumstances in which membranes processes become periodic on a macroscopic scale. This may in principle arise from the coupling between several flows crossing simultaneously the same membrane, or as a result of periodic chemical reactions. Under certain conditions coupling between diffusional and volume flows and electric current were demonstrated both, experimentally (Teorell, 1955) and theoretically (Schlögl, 1964) to give rise to oscillations quite similar to those described in excitable tissues. Self-sustained oscillations in chemical reactions were also demonstrated *in vitro* (Chance, Ghosh, Higgins and Maitra, 1964) and the necessary and sufficient conditions analyzed theoretically (Spangler and Snell, 1967; Katshalsky and Spangler, 1968). Feedback coupling between two enzymatic processes presents under certain conditions an activity-concentration curve characterized by two metastable states (Spangler and Snell, 1967). When the intensive parameter (concentration) crosses the transition region (fig. 8–3) from left to right, the system surpasses transition line and becomes metastable until a critical point is reached in which it suddenly switches to another stable state in which it has a higher enzymatic activity. An equivalent phenomena is observed in going from right to left. The binding of ligands to a two-dimensional array of subunits, in which the subunits can adopt two different metastable states, and are subject to nearest neighbour interactions, can also show that, at a given concentration of ligands, the

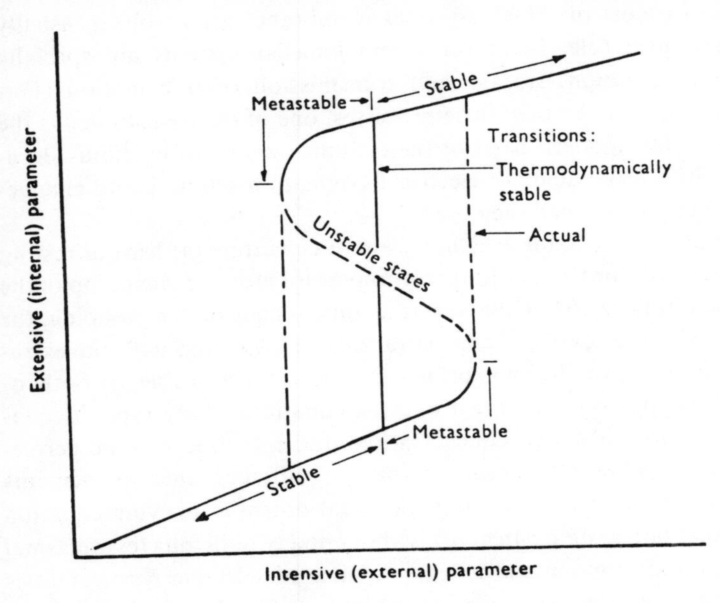

FIGURE 8–3 Schematic representation of general face transition, depicting regions of stability, metastability and unstability. Ordinate: extensive parameter (*e.g.* flow, current). Abscissa: intensive parameter (*e.g.* concentration, voltage). (Taken with kind permission from A. Katchalsky and R. A. Spangler. 1968.)

system switch from a metastable state to the other (Changeux, Thiery, Tung, and Kittel, 1967). These subunits can be enzymes catalyzing a reaction. Katchalsky and Spangler (1968) have shown that if the agent inducing the transition is a reactant or a product of the reaction being catalysed, oscillatory behaviour can be achieved by appropriately coupling the reaction with a membrane transport process regulating the flow of reactant or product. In other words: a membrane with allosteric subunits can show oscillatory behaviour as a consequence of the coupling of the chemical processes in which the subunits participate, with the membrane transport. Oscillatory behaviour are the quintessence of biological rhythms, of the transmission and processing of nervous signals, of the mechanochemical reactions in muscle, etc. One of the non-linear behaviours best studied in membranes is the firing of the action potential.

Cells respond to electrical, mechanical, photic, thermal and chemical stimulations. In most of these responses membranes are involved, usually playing the central role. Membranes and lamellar systems are specially suited for the reception, storage and transmission of information (Fernández Morán, 1962). Among those processes, one of the best studied is the nerve impulse. The modern basis of these studies were laid by Julius Bernstein (1912), who associated the electrical events to a selective and changeable permeability of the cell membrane.

The ion pump of the axon does not seem to constitute (at least at resting physiological condition) a quantitatively important electrogenic pump of the type described in page 154. However as a consequence of the pumping, the cell has high K^+ and low Na^+ concentrations as compared with the extracellular solution. Since the membrane is passively permeable to both of them, they move down the gradient obeying equations of the type of equation 3–3. Since not only their concentration gradients, but also the permeability of the membrane for each of them is different, their movements across the membrane gives rise to an electrical potential obeying equation 6–25. This potential *(resting potential)* is about ninety millivolts (cytoplasmal side negative) and represents a delicate balance of selective permeabilities and concentrations gradients. Concentrations (particularly in large cells with low surface-to-volume ratio) do not change appreciably during nerve activity. The major variables are thus potentials and specific permeabilities. The interrelationship between these variables was studied using a *voltage clamp* technique. An electrode is introduced into the axon to measure the voltage across the membrane using a reference electrode placed in the extracellular solution. Another pair of electrodes, one inside and the other outside the axon, are used to pass current across the membrane. The electronic device is build in such a way that by passing current in one direction or the other it maintains the voltage across the membrane at any desired level (Cole, 1949; Hodgkin, Huxley and Katz, 1949). Figure 8–4 shows the time course of the Na^+ permeability (center) and K^+ permeability (bottom) expressed as conductances, as the voltage (upper trace) is clamped 56 mV above the base level (Hodgkin, 1958). There is a sudden and transient increase in Na^+ permeability. Na^+ being more concentrated in the extracellular than in the intracellular solution, tends to penetrate in the cell. Should the voltage of the axon being not clamped by the electronic device, this entrance of Na^+ would had made the cytoplasm positive with respect to the outside. This is a transient increase which gives way to a decrease in the Na^+ conductance, and increase in K^+ conductance. Since the concentration of K^+ is higher in the cells than in the bathing solution, it tends to leave the cell.

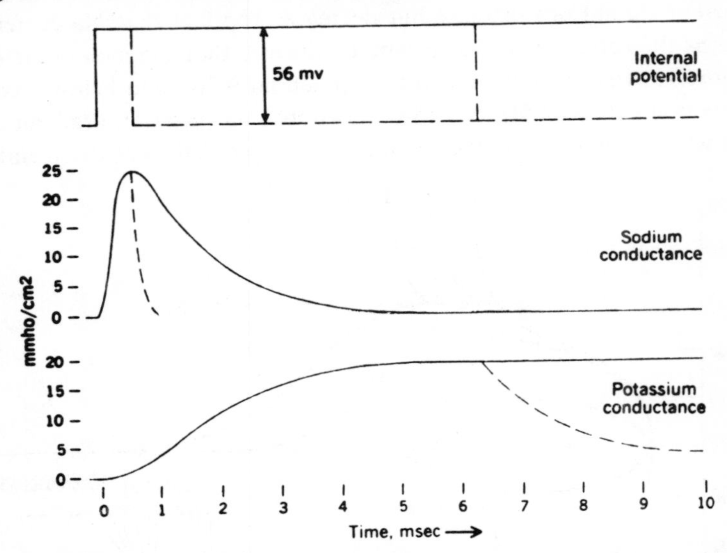

FIGURE 8–4 Time course of the sodium and potassium conductances associated with a depolarization of 56 mV. The continuous curves are for a maintained voltage step. The broken curves show the effect of repolarizing the membrane after two intervals indicated by broken vertical lines in the trace at the top. (Taken with kind permission from A. L. Hodgkin, 1958.)

The exit of K^+ tends to make the cytoplasmal side negative, but again, the clamping keeps the voltage unchanged. Doted lines show what happens when the applied voltage is interrupted. If instead of changing 56 mV, a different value is chosen, the relative values of the Na^+ and K^+ conductances achieve different values than those in figure 8–4. One of the main teachings of these experiences is that the extent to which the membrane selects K^+ or Na^+ depends on the magnitude and sign of its potential which, in turn, depends on the selectivity.

How are actually permeability and voltage changes associated in a non-clamped axon? In other words: how does the nerve impulse occur? If a square pulse of current is passed inward in an axon having a resting potential of some -80 mV, the potential becomes more negative (figure 8–5). If the pulse is passed outwards it depolarizes the membrane (tends to cancel the membrane potential). According to figure 8–11 as the potential in the cell is made less negative, the Na-conductance increases, and the Na^+ current also increases. According Ohm's law, when the electrical potential decreases the current should not increase but decrease. The fact that Na current *increases* as the cell is made *less* negative indicates that the membrane offers a *negative resistance*. If the membrane potential is lowered below a certain unstable point (threshold), Na flows down its concentration gradient along a path which become selective for this ion and off balances the membrane

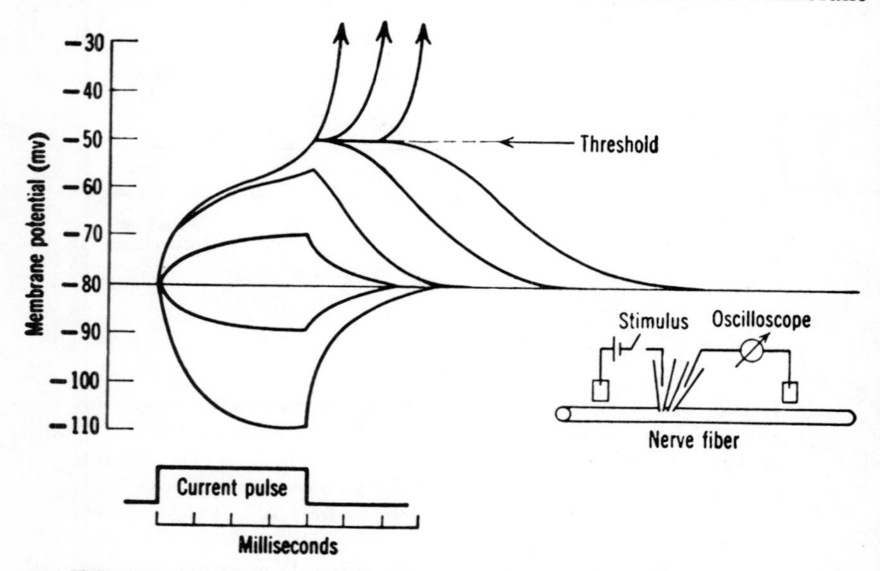

FIGURE 8–5 Changes of the local potential across an axon membrane due to rectangular pulses. (Taken with kind permission from B. Katz, 1966.)

potential or may even reverse it (*i.e.* make the cytoplasm positive). But figure 8–4 shows that the increase in Na-conductance is a transient phenomenon. As the Na conductance comes back to its low, pre-stimulation levels, it is followed by an increase in K-conductance. This permits a larger amount of K^+ to flow outwards. Its electrical consequence is the increase in the negativity of the cytoplasm.

This increase of K-conductance is a process of negative feedback and eventually cancels off permiting the fibre to recover the K/Na selectivity of its resting state. This change of the membrane potential from its resting negative value, to a positive and then back to a negative one, occurs in a few milliseconds and is called action potential or "spike". In some axons the amount of K^+ displaced during an action potential is about one millionth of the K^+ contained in the axon. Therefore the axon can fire a large number of action potentials before its K^+ content is appreciably altered. Nevertheless the operation of the Na–K-pump keeps the ion balance in steady state.

The electrical spike, characteristic of membrane excitation in a given point, propagates to neighbour areas of the membrane. If the axon behaved as a common cable, the excitation would become smaller as it travels to neighbour areas until it vanished at a relatively short distance from the point of origin. The fact that the Na^+ entrance following an all-or-none response, produces a potential much in excess of the stimulation threshold, insures that the neighbour area will receive an effective signal despite of the dissipation produced by the cable structure. Hence the signal travels throughout the membrane.

This is a very schematic description of the events which may be observed not only in axons but in other excitable cells as well. For key papers and reviews see: Hodgkin and Huxley (1939), Cole and Curtis (1939), Hodgkin and Katz (1949), Keynes and Lewis (1951), Hodgkin and Huxley (1952), Hodgkin (1958), Huxley (1964) and Katz (1966). The view expressed here though, is not unanimously accepted. For other points of view see: Tasaki and Takenaka (1964), Tasaki and Singer (1966), Troshin (1966) and Ling (1969).

Electrical signals and ionic movements are not the only manifestation of excitability. Hill (1932) described an "initial nerve heat" during or immediately after the passage of an electrical spike. This was taken as an indication of a chemical reaction associated with the impulse. Optical properties (absorption, fluorescence, turbidity, birefringence, and light scattering) change during stimulation, suggesting a rapid reorientation of the membrane molecules (Ungar, Ascheim, Psychoyos and Romano, 1957; Ungar and Romano, 1962; Tasaki, Watanabe, Sandblon and Carnay, 1968; Cohen, Keynes, and Hill, 1968). The pore radius of the axolema at rest is about 4.7 Å. It switchs to 6.2 Å on stimulation at 100 impulses per second (Villegas, Bruzual and Villegas, 1968). This also suggests a structural change underlaying the electrical events. Goldman (1964) suggested that the polar groups of the phospholipids by changing their position at the interfase membrane-solution could act as gates, of the channels where Na^+ and K^+ diffuse. Fischer, Cellino, Gariglio and Tellez–Nagel (1968) have observed that the protein metabolism of the squid axon is enhanced during electrical stimulation.

As discussed in chapter 1 the participation of water in the structure of the membrane and its state in the vicinity of the membrane are likely to bear an enormous influence on membrane phenomena. The state of water also changes during the electrical depolarization of the membrane as indicated by nuclear magnetic resonance analysis (Fritz and Swift, 1967).

There are two points worth noticing before we leave this topic: 1) Membranes are not homogeneous all over their area. They have specialized spots, like synapsis and receptors, specially suited to receive information. This is a very wide field that will not be considered here. 2) Electrical all-or-none responses, although highly developed in neurons and muscle cells, can be also elicited not only in other cell membranes, but in complex epithelial membranes as well (Finkelstein, 1964; Bueno and Corchs, 1968; Lindeman, 1968). We have seen in page 145 that even artificial membranes can be prepared to fire action potentials.

8.4 The effect of hormones

Many hormones exert profound influence on membrane phenomena either by acting directly on the membrane or through some intermediate reaction. The resulting membrane change can, in turn, modify other cellular functions in which membranes are involved. Hormones, can for instance, act at the membrane level and modify the metabolism of the cell through, in principle, three different mechanisms.

1) *By changing the specific permeability to certain substances.* As mentioned in the Introduction, the course of a metabolic reaction may depend on the rate at which substrates and products enter or leave the compartment where the reaction takes place. Therefore, hormones and other substances may control chemical events by modifying the permeability of membranes for specific substances. Already in 1914, long before the discovery of insulin, Höber suggested that diabetes might be due to an alteration of sugar entrance into the cell compartment. Lundergaard, Levine and others have later shown that insulin increases the cellular uptake of sugars (Levine, 1961).

2) *By changing the reactivity of molecules that are part of the membrane structure:* as discussed in chapter 1, water is an important component of the membrane structure, and its organization can change the mobility of protons

by orders of magnitude. In connection with this, it is very interesting to notice that certain hormones, in particular the steroid ones, rank among the best ice structure-promoters known (Fukuta and Mason, 1963). Also any effect that the hormones exert on the lipid and protein component is likely to be propagated to other areas in the membranes as discussed in page 46. The behaviour of enzymes attached or included in these structures might thus be altered.

3) *By changing the activity of enzymes located in structural elements inside the cell:* Peters (1956; see also Hetcher and Lester 1960 and chapter 9) has suggested that the cell has a cytoskelton which would act as a sort of framework for enzymes. The cytoskelton, and thereby the activity of the attached enzymes, would be changed as a consequence of the binding of a hormone to a specific receptor in the membrane

We will illustrate the influence of hormones on membrane phenomena by discussing the effect of the antidiuretic hormone.

8.4.1 The antidiuretic hormone (ADH)

One of the reasons animals keep their internal environment constant in a changing external environment, is their ability to regulate the exchange of substances across their epithelial membranes. Animals respond to a water deficiency produced by a lack of drinking or by an excess of swet by secreting ADH. This hormone acts on membranes like the renal tubule or the frog skin producing an increase in the water uptake which tends to preserve the water content. Classic studies carried out by Ussing and Zerahn (1951) Koefoed–Johnsen and Ussing (1953, 1958) and Andersen and Ussing (1957) oriented the research toward the modifications that ADH produces at the membrane level. The information collected there since indicates that ADH has roughly three main effects:

1) *ADH effect on Na transport:* As revealed by tracer kinetics and short circuit current measurements, ADH increases the Na^+ transport. This effect was interpreted on the basis of transcellular models as an increase in the Na-permeability of the outer barrier (Leaf and Dempsey, 1960; Frazier, Dempsey and Leaf, 1962; Curran, Herrera and Flanigan, 1963). The basis for this view are as follows: a) as the Na^+ concentration of the outer barrier side is increased the net Na^+ influx across an epithelial membrane increases to a maximum and then saturates. This could be due to saturation of any of the several steps in the Na^+ influx. b) the Na "pool" does not exceed a certain maximum value, indicating that the flux outer solution→pool saturates.

c) A plot of the flux pool→inner solution *vs.* amount of Na in the pool does not reach saturation. Hence the limiting step seems to be the penetration across the outer barrier (see Kirschner, 1955; Snell and Leeman, 1957; and Morel, 1958; Cereijido, Herrera, Flanigan and Curran, 1964). ADH would increase the rate of this step by making the outer barrier more permeable. The decrease of the resistance of the outer barrier that would accompany such a change was measured by Whittembury (1962) and Civan and Frazier (1968) using a microelectrode technique.

In the methods of tracer kinetics used, the flux across the outer barrier was estimated indirectly and relies on the validity of several assumptions. Cereijido, Rotunno and Vilallonga (1969) have used a technique for the *direct* measurement of the flux outer solution→epithelium. According to table 8–1 the flux outer solution→epithelium is larger than the total influx. This suggests that, if Na^+ influx consists of two steps, the second one should be rate

TABLE 8-1[a] Na fluxes in frog skin

	Total influx	Flux outer solution – epithelium
	µmole/hour cm²	
L. ocellatus	4.2 ± 0.3	8.9 ± 1.4
R. pipiens	1.4[b]	5.9 ± 0.6

[a] From Cereijido, Rotunno and Vilallonga (1969)]
[b] [From Curran, Herrera and Flanigan (1963)]

limiting. Cereijido, Migliora and Rotunno (1969) using the technique for the direct measurement of the flux outer solution→epithelium have found that *both* fluxes are increased by ADH. According to non-transcellular models (see page 169) there are two ways in which these experimental results can be explained: a) The increase in permeability of the outer barrier allows Na^+ to penetrate into the cell and be expelled via transcellular routes. In this way more pumps would be recruited. b) The ADH might have a direct influence over the pumping system. Both of these possibilities can be given simultaneously.

Hong, Park, Park and Kim (1968) demonstrated that energetically depleted frog skins will only respond to ADH if a metabolic substrate like piruvate is added. This reflects the fact that either by offering more Na to the pump or by direct stimulation, the ADH increases the active step. It is even possible that ADH action is not excerted directly on the two steps but were mediated by Adenosine 3′5′-monophosphate (cyclic-AMP) (Orloff and Handler, 1963; Schwartz and Walter, 1967; see also Sutherland and Rale, 1969).

2) *ADH effect on the flux of small molecules.* Andersen and Ussing (1957) have found that ADH increases the permeability of the frog skin to thiourea and acetamide. It also elicits the same effect in other tissues. Leaf and Hays (1962) have found that of some 40 molecular species tested, ADH only increased the penetration of certain small and uncharged amides and alcohols.

3) *ADH effect on water movement.* ADH increases the permeability to water of the frog skin and other epithelial membranes. If an osmotic gradient is applied across the membrane (*e.g.* by making the outside solution hypotonic) this increase in permeability results in a large water influx. From the facts that osmotic gradients operate through water pores, and that ADH permits a larger amount of water to flow under a given osmotic gradient, it is deduced that ADH must increase either the population and/or the size of the pores. It was postulated that these pores are located on the outer barrier since this barrier is by far the main restriction to water movement. It was also observed that an epithelium in contact with a hypotonic solution on the outside will only swell if ADH is added (Koefoed–Johnsen and Ussing, 1953; MacRobbie and Ussing, 1961; Hays and Leaf, 1962b; Whittembury, 1962).

The effect of ADH on water movement does not depend necessarily on the active transport of Na^+ (Hays and Leaf, 1962a; Bourget and Morel, 1967). However in practice—or better, in the physiological state—Na or NaCl pumping builds up the osmotic gradient necessary for water movement (see page 182).

For other examples of hormones acting directly or indirectly on membranes see: Herrera, 1965; Crabbé and deWeer, 1965; Reiss and Kipnis, 1959; Kostyo and Engel, 1960; Christensen, 1962; Zadunaisky, Parisi and Montoreano, 1963; Edelman, Bogoroch and Porter, 1963.

The examples discussed in this chapter of membranes that can incorporate a new translocating mechanism (permeases); remove a whole piece to form a pinosome (pinocytosis); change and restore their ion selectivity (nerve impulse); and vary the size of the pores to serve the functional needs of the whole animal (ADH effect), were chosen to illustrate the fact that membranes are not static barriers but highly sensitive and versatile molecular organizations.

9

The association-induction hypothesis

PERHAPS THE most condensed summary of the preceding chapters is: the cell membrane determines the rate of selective entry and exit of water and solutes into and out of the cell as well as their concentration levels. We may refer to this view as the "Membrane Theory". In this chapter we shall consider an almost completely different point of view whose summary might be: the steady levels of solutes and water do not represent steady states maintained by the cell membrane, instead the steady levels represent equilibrium states which reflect the properties of the whole protoplasm. We will first review some of the critics to the Membrane Theory. Then we will discuss Ling's Association-Induction Hypothesis which is one of the most interesting alternatives to Membrane Theory. In order to understand this position we will have to introduce some new information (induction, *c*-value, cardinal site, gang, etc.).

9.1 Critics to the membrane theory

The membrane picture which is most heavily criticized is that of a static lipid bilayer full of pores, specific carriers and pumps, and which separates two well-stirred solutions, one of them being the protoplasm. Let us review some of the most acute critics to the fundamental assumptions of the Membrane Theory.

9.1.1 The energy requirement

As discussed in chapter 3 and 5, the cell maintains an asymmetric steady state distribution of solutes which the Membrane theory associates to pumping mechanisms. Using values of the rate of ion pumping, ion permeabilities, and electrical potentials in muscle cells published by several authors, Ling (1965) has calculated that frog muscle would consume: 51 cal/kg per hour in Na^+ pumping (data from Levi and Ussing, 1948; Harris and Burns, 1949; and Keynes and Maisel, 1954), 343 cal/kg per hour in Ca^{++} pumping (data from Shanes and Bianchi, 1959; and Tasaki, Teorell and Spiropoulos, 1961), 176 cal/kg hour in Mg^{++} pumping (data from Conway and Cruess–Callaghan, 1937). Therefore, the pumping of these three ions alone would require 570 cal/kg hour. According to Conway (1946) the resting energy output for frog muscles is about 170 cal/kg per hour. Thus, the energy needed for these three pumps, even if they operated with 100% efficiency, would amount to 300% of the total energy output of the resting muscle cell. As Ling puts it: "unless we are willing to venture that the second law of thermodynamics does not hold in these living cells... there is no alternative to discarding the pump mechanism for selective ionic distribution in living cells". This author makes it clear that his conclusions have no direct bearing on true active transport across epithelial membranes of the type we have studied in chapter 7.

9.1.2 The state of water and ions in the cell

In Membrane Theory intracellular water is generally assumed to be in free state and behave as a continuous dielectric; its sole influence on protein behaviour is exercised through its macroscopic dielectric constant. If this were true, it would not make any difference for a biological system the replacement of its water by deuterium oxide. Deuterium oxide, though, has profound effects not only in cells and multicellular organisms (Thompson, 1966) but also in isolated proteins (Kritchevsky, 1960). This would indicate that the almost identical macroscopical properties of water and deuterium oxide are not enough to insure an equal functioning of biological systems, and that minute molecular details of the water molecules should be considered. Besides, the freezing pattern of water in living cells (Ling 1969a) as well as its nuclear magnetic resonance spectra (Chapman and McLaughlan, 1967; Fritz and Swift, 1967) indicate that the state of water in the cell is not the one it would have in a diluted salt solution.

Ions in the cytoplasm do not seem to be in the free dissociated form either as reflected by Na^+ fixation in isolated actomyosin (Lewis and Saroff, 1957), the chemical activity of intracellular Na^+ (Hinke, 1959, Ling, 1969b) and the nuclear magnetic resonance spectra of Na^+ in several tissues (Cope, 1967; Rotunno, Kowalewski and Cereijido, 1967).

9.1.3 Influx profile analysis

The penetration of water into single frog ovarian eggs follows a time course which may be taken to indicate that the cytoplasm offers the same resistance to the movement of water as does the cell membrane (Ling, Ochsenfeld and Karreman, 1967). This in turn means that intracellular water is not entirely in liquid state.

9.1.4 The behaviour of proteins

In the opinion of the critics of the Membrane Theory, part of the trouble comes from the old Linderstrøm–Lang's model of proteins, which is still used in devising models of the protoplasm and which envisages a protein as if it were a spheroid with a net charge smeared over its surface. Net charge would only be determined by the pH of the medium and the number of aminoacid residues that carry ionic side chains. K^+ and Na^+ would only play a role as part of the ionic strength and, therefore would be indistinguishable. However the enormous amount of information stored in a protein, the tremendous specificity of enzymes, the marked effect on enzyme activity exerted by minute amounts of certain ions, etc. shows that the Linderstrøm–Lang model of a protein is—at best—an oversimplification.

If these critics were correct, the Membrane Theory would be untenable. To explain the movement and distribution of ions without relaying in the assumptions made in Membrane Theory two main models were put forward: *The Sorption Theory*, favoured mainly by the Soviet School (for a recent excellent review see Troshin, 1966) and the *Association-Induction Hypothesis* presented by G.N.Ling. Both of them have many points in common. We shall base our illustration of "non-membrane" models by sketching Ling's hypothesis. For a more thorough discussion of this hypothesis see: Ling, 1960, 1962, 1965, 1967, 1969a and b; Ling and Ochsenfeld, 1965, 1966 and 1968; and Ling, Ochsenfeld and Karreman, 1967. Although we shall skip the historic development of the idea that a cell is not just a droplet of water in which enzymes, substrates, etc. mix up as in a soup, and will focus our attention in its present state, it is opportune to indicate that the idea is even older than the Membrane Theory itself and was suggested by Dujardin (1835) and Mohl (1846).

9.2 *The inductive effect*

This concept was formally introduced by Lewis (1923) to characterize the change in reactivity of a chemical group produced by the introduction in the molecule of another chemical group which, as we shall see below, may be located at a considerable distance from the first. Figure 9–1 depicts a molecule of acetic acid in which the dots represent the electronic cloud. When the hydrogen atoms on the methyl group are replaced by chlorine, the electrons

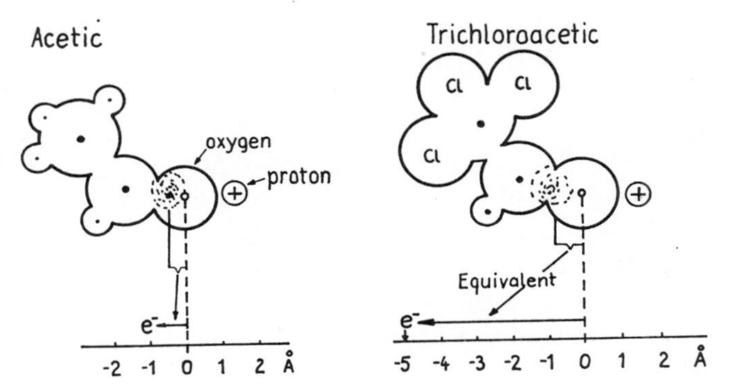

FIGURE 9–1 The *c*-value. See text.

of the molecule are "sucked" toward the chlorine atoms. The electron density at the carboxyl group is reduced with the consequence that the attraction between this group and the proton is weakened. The fact that trichloroacetic acid has three chlorine atoms where the acetic acid has three hydrogens lowers its attraction for protons "at the other end" of the molecule. This is reflected by the fact that the pK of the acid is 4.76 while the pK of the trichloroacetic acid is lower than 1. Chlorine *induces* a change in the affinity for protons in the carboxyl group. The extent to which this effect is transmitted is greater the greater the polarizability of the intervening bonds. Acetophenoneoxime, for instance, dimerizes by hydrogen bond formation involving the NOH group. The fact that the free energy of dimerization of

$$X-C_6H_4-C(CH_3)=NOH$$

acetophenoneoxime

this compound is altered by the nature of the para-substituent (Reiser, 1959) indicates that the hydrogen bonding can also be inductively affected. The ability to induce a given change is so important and predictable that an inductive constant was introduced to characterize the inductive ability of different substituents (Hammett, 1940; Taft, 1953).

9.3 c-value

In order to characterize the proton affinity of an acidic group one uses the pK value. However, since inductive effects are exerted over a variety of functional groups other than those involved in proton dissociation, one needs a new parameter to characterize the state of the involved groups. Since the phenomenon underlaying all these effects is a displacement of the electron cloud towards or away of the group concerned (fig.9–1), the new parameter *(c-value)* is expressed as a function of the number of Ångstrom units that the center of the electron cloud has been displaced. In fact it is the distance that *a single* negative charge would have to be displaced to produce an electrostatic change equivalent to the one produced by the displacement of the whole electronic cloud (see figure 9–1). The *c*-value is positive when the displacement is towards the functional group and negative when it goes toward the molecule to which the functional group is attached. A rigorous definition and treatment of this parameter would take too long a digression, and will involve concepts not considered in this book. The consult of Dr. Ling's original work (1962) is recommended.

An acidic group like acetic acid (pK 4.76) has a *c*-value of around -1 Å. The minus sign indicates that the electron cloud escapes from the group and toward the rest of the molecule. Trichloroacetic acid (pK < 1) has a *c*-value of around -5 Å. An analogue c'-value can be calculated for basic groups. This time it is expressed as the number of Ångstroms that a *positive* charge has to be displaced. The use of c (or c') values is not restricted to net charges but may be also applied to any group, like the proton-accepting C=O group or the proton-donating NH group.

9.4 c-values and ion selectivity

In the case of an anionic group the *c*-value is roughly an expression of its anionic strength. One can take an acidic group of a given *c*-value (say -5 Å) and calculate the energy of its interaction with Na^+. Then assume that a water molecule is interposed between the anionic group and the Na^+ ion, and calculate the energy of the anion-water-Na^+ assembly. This time of course, the interactions of the fixed anion and the mobile cation with the permanent and induced dipoles of water molecule should also be considered. The same calculation may be repeated for the case of 2 or 3 molecules of water interposed, this time including also the interactions of the water molecules among themselves. This, in turn, permits to calculate the statistical

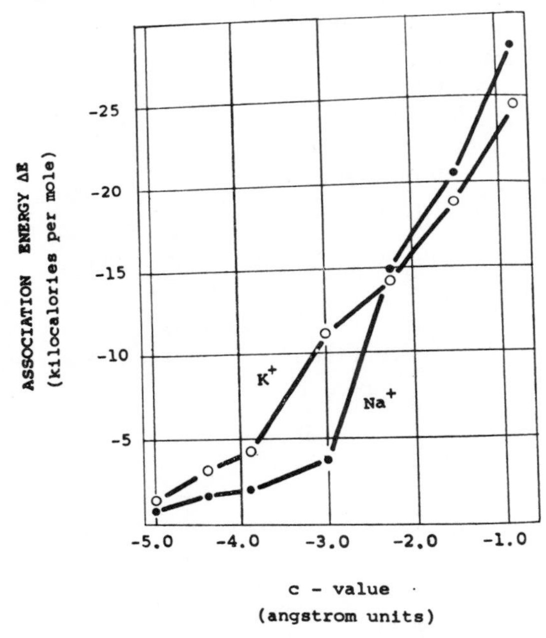

FIGURE 9–2 Theoretical curves showing ion selectivity as a function of the *c*-values. In this chart, and those in figure 9–4 the ion prefered at a given *c*-value is the one occupying the upper position. (Taken with kind permission from G.N.Ling, 1962.)

weight of each configuration (*i.e.*, with 0, 1, 2 or 3 water molecules) and the average association energy of Na^+ with the fixed charge having a *c*-value of -5 Å. This association energy is then plotted in a graph as a function of the *c*-value (fig. 9–2). If the whole calculation is repeated with an anionic group of a *c*-value of -4, -2, etc., an isotherm for Na will be obtained. Now we can do all over for K^+ instead of Na^+. Notice that the isotherms cross each other at a given *c*-value. This shows that the selectivity for cations is a function of the *c*-value of the group. The orders of selectivity found for Rb^+, Cs^+, K^+, Na^+, and Li^+ coincide with those discussed in chapter 6. This is not surprising since, as stated above, in the case of an anionic group, the *c*-value is a measure of its anionic strength.

9.5 *The inductive effect in proteins*

Having studied in chapter 5 that a few millimols of Mg^{++} change reversibly the behaviour of the ATPase, and that the rate at which this enzyme breaks down ATP is strongly dependent on the nature and concentration of the cations present, we hardly need more examples of the specificity—and reversible changes in specificity—that a protein may exhibit. The fact that the combination of an oxygen atom with an heme group of the hemoglobin *increases* the affinity for oxygen of the other hemes—located in separated peptide chains—indicates that the information that an oxygen has been attached to the first heme is transmitted throughout the protein molecule. Since classical models, picturing proteins as spheroids with smeared charges are not adequated to describe this effect (as well as many other properties of the protein), Ling developed a model of the protein molecule that would account for those physicochemical properties and permit, in particular, the transmission of inductive effects.

Proteins are linear polymers containing hundreds of units of some twenty-three aminoacid species. Chemical analysis reveals that the aminoacid sequency of each protein is unique, and that a minor alteration (like the change of a histidine residue in the fifty-eighth position of the alpha chain or the sixty-third position of the beta chain of the hemoglobin molecule by a tyrosine residue) suffices to modify its properties. Two types of molecular groups are important for the inductive effect: a) *Peptide linkages* which transmit very easily the inductive effect thanks to the high polarizability of the peptide amide bond due to its partial resonance (Pauling, 1960). b) *The saturated*

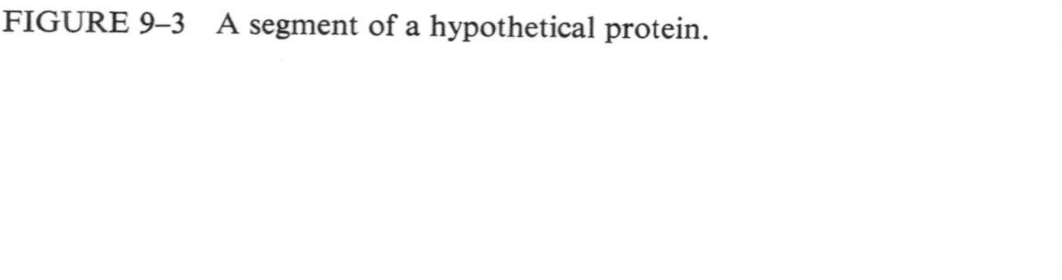

FIGURE 9–3 A segment of a hypothetical protein.

carbon atoms (CH_2) that separate the side chain functional groups from the polypeptide chains (fig. 9–3) and which offer a relatively high resistance to the propagation of the inductive effect. Serine, threonine, cysteine, cystine, aspartic acid, asparagine, phenylalanine, tyrosine, histidine, tryptophane, proline and hydroxyproline functional groups are separated from the polypeptide chain only by one saturated carbon atom. Peptide linkages and CH_2 groups act as conductors and resistors in the network where the inductive influence propagates. Due to the inductive effect, the *c*-value (or *c'*-value) of a functional group in a protein will be determined not only by the nature of the particular aminoacid residue, but also by the nature of the immediately adjacent groups and by the counterions or H-bonding partners of the adjacent groups. Thus the dissociation constant of a negative group is not influenced by another negative group down the chain in the same manner when a positive group is interposed in the sequency than when another negative group is intercalated, or when it is separated from the polypeptide chain by one than by three CH_2 groups. The total electronic profile of the entire protein molecule is determined by its sequency of aminoacids and its environment.

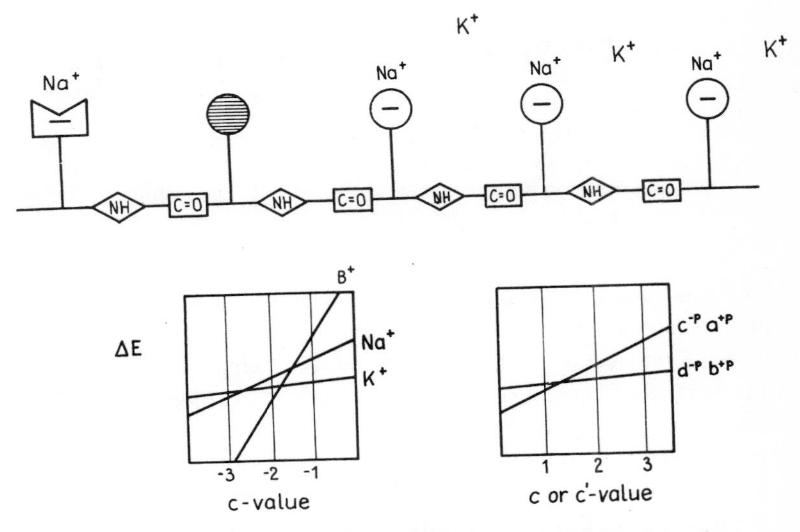

FIGURE 9–4 Theoretical model of cooperative transformation of a protein chain. The site on the left is a cardinal site. The charts show in a simplified way the relationship between association energy and the *c*-value. (After G.N. Ling, 1962.)

An example of cooperative transformation. Figure 9–4 illustrates a protein segment with Na^+ fixed to its anionic groups. The anionic group on the left hand side has a c-value of -1 and selectivity $B^+ > Na^+ > K^+$ (see enclosed graph). B^+ is an unspecified cation which is not present at the moment in the system. Since B^+ is not present the site binds Na^+. The other anionic groups have a c-value of -2 (selectivity $Na^+ > K^+ > B^+$). a^{+p} and b^{+p} are proton donating molecules. c^{-p} and d^{-p} are proton accepting molecules. C=O are proton accepting groups of the polypeptide chain with a c-value of 1, (selectivity $b^{+p} > a^{+p}$). NH are proton donating groups with a c'-value of 1 (selectivity $d^{-p} > c^{-p}$). The graph indicates the affinities as a function of c or c'. If B^+ is now introduced into the system, it will displace Na^+ from the first site on the left (fig. 9–5). But the state of this site is not the same when it has B^+ than when is has Na^+, just as the acid in figure 9–1 was not the same when it had chlorides replacing the hydrogens. The change induced

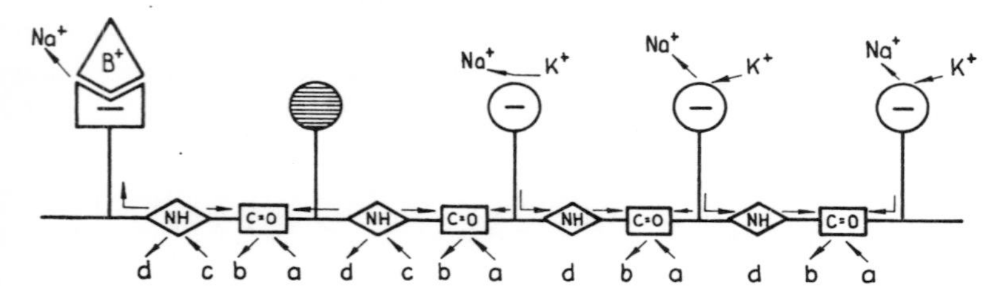

FIGURE 9–5 This figure shows the same theoretical model as in figure 9–4 this time it shows the effect of introducing the *cardinal absorbant* B^+. The thin arrows on the polypeptide chain indicate the direction of electron displacement. (After G.N.Ling, 1962.)

by B^+ produces a rearrangement of the electrons (thin arrows) that bring about an increase in the c'-value of the neighbor NH group from 1 to 2 (hence c^{-p} replaces d^{-p}). This in turn induces a change (from c-value 1 to 2) in the next C=O group (hence a^{+p} replaces b^{+p}). Electrons in the anionic groups are displaced toward the peptide chain and the c-value of these groups changes from -2 to -3. According to the affinity graph, when the sites have a c-value of -3, they prefer K^+ over Na^+. Therefore, as a consequence of the introduction of B^+ the state of the protein was changed from a state of Na^+ affinity to one of K^+ affinity (fig. 9–6). In this particular example only ions and small molecules interact with the protein. However there is no reason to restrict the model to such interactions. The groups could also form intramolecular H-bonds, salt linkages or interact with a whole conglomerate of proteins (fig. 9–7). b^{+p}, for instance, could be a functional group belonging to another part of the same protein or to another protein, an aminoacid, a sugar, etc. Therefore the change in *c*-value may result in a

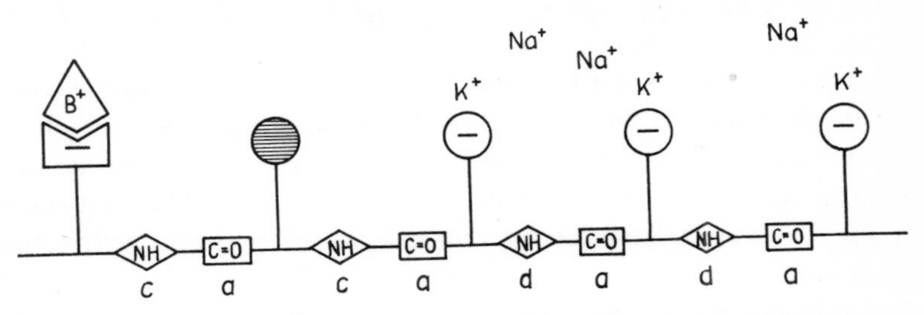

FIGURE 9–6 This figure shows also the same theoretical model as in figure 9–4 and 9–5. As a consequence of the introduction of the cardinal absorbent *B* the polypeptide achieves a new metastable state in which it has selectivity for K^+.

change of affinity, adsorption, unfolding, dissociation etc. As a matter of fact, since unicellular organisms, like individual cells of multicellular organisms, do behave in coordinated and purposeful fashion, the Association-Induction Hypothesis maintains that proteins in the cell interact with each other even if they are not linked together by covalent bonds. In this case, the inductive effect elicited in a given site *(cardinal site)* of a protein can be

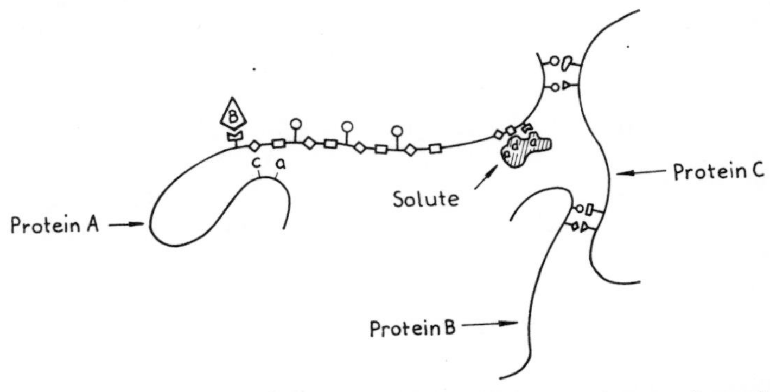

FIGURE 9–7 Diagramatic representation of a cooperative transformation induced by a cardinal absorbant. The participating groups may belong to the same or different polypeptide chains, to other solutes, etc. (After G. N. Ling, 1962.)

propagated to sites located in a relatively extense portion of protoplasm. The groups of sites responding to a change in a cardinal site are said to form its *gang*.

9.6 The state of water and solutes inside the cell

According to the Association-Induction Hypothesis the other factor that plays an outstanding role in selectivity as well as many other cellular events, is water. Therefore we shall now consider some aspects of the state of water in the cell to see what sort of influence it may exert. At this point the reader may find useful to refresh the few concepts on water properties discussed in chapter 1.

A survey carried out by Ling (1962) of the aminoacid composition of a wide variety of cells, ranging from bacteria to muscle cells indicated that the concentration of ionic side chains in cellular water is no less than 190 mM. Let us see what does it mean in terms of spatial distribution of charges. Cells have some 17.8 per cent of protein with an average molecular weight of about 112 per aminoacid residues. Therefore, assuming that the specific gravity of the cells is 1.05, the concentration of aminoacid residues in the cells is 1.66 moles per liter of fresh cells. By multiplying by Avogadro's number we obtain

$$1.66 \times 6.06 \times 10^{23} = 1.006 \times 10^{24}$$

that is, the number of aminoacid residues in a liter of fresh cells. Each peptide linkage measures 3.5 Å. If each aminoacid residue were put along a fully extended chain, the length of this chain would be

$$1.006 \times 10^{24} \times 3.5 \times 10^{-8} = 3.52 \times 10^{16} \text{ cm}$$

or 3.52×10^{15} filaments measuring 10 cm in length. If we take a square surface of 10×10 cm and nail the filaments so that they will stand uniformly distributed, there would be

$$\sqrt{3.52 \times 10^{15}} = 5.93 \times 10^{7}$$

filaments on each side, separated from each other by

$$\frac{10}{5.93 \times 10^{7}} = 16.9 \times 10^{-8} \text{ cm}.$$

In other words, the filaments will stand 16.9 Å apart. On the other hand, some 8 per cent of the total aminoacids have a free cationic group (carried by lysine + arginine + ionized histidine) and some 10 per cent have a free anionic group (glutamic + aspartic). This implies that 18 per cent of the aminoacid residues have a charge. If we distribute them homogeneously along the filaments there would be a charge each 5.55 peptide linkages, or every $5.55 \times 3.5 = 19.4$ Å (let us say, roughly, 20 Å). Since chain-to-chain distance and the charge-to-charge distance on the filaments are, at most, 20 Å, Ling drew a diagram in which each charge occupies a point at some 20 Å from its neighbors. Each charge occupies the center of a microcell in which ions and water are influenced and do not behave as in free solution, (fig. 9–8). The total volume of the microcells takes most of the cell volume. There is therefore little chance that ions and water will not be subject to the influence of the fixed charges. Even when the protein in a cell are not arranged as in figure 9–8, it is interesting to notice that a cell endowed with such a large number of charges constitutes a fixed charge system. Besides, the cells have a concentration of hydrogen bonding sites around 3.5 to 7 M afforded by the peptide amide groups and side chains of the proteins which are likely to influence the behaviour of water in the cell. Water in the cell is therefore likely to be polarized and hydrogen bonded (Ling, 1962, and Ling, Ochsenfeld and Karreman, 1967) as it is around collagen (Berendsen, 1962) and DNA molecules (Hearst and Vinograd, 1961). As discussed before, the idea that the state of water in biological systems might not be as in free solution, is not new in Biology. What the Association-Induction Hypothesis emphasizes is the concept that the non-liquid state of water in the cell has a more profound effect than to produce a mere deviation from "ideality" of the intracellular solutions.

Since the number of water molecules that participate in the ion-fixed charge association plays a role in ion selectivity, the size of the microcells, and the number of water molecules that they can accommodate is likely to be important in determining the nature of the ion preferred in the lattice. We have also mentioned in page 17 that one of the reasons ions dissociate in aqueous solution is the rotational entropy gained by the hydrated ions. If water inside the cell is polarized and not free to move, the entropy gained by ions upon dissociation will be very low. In other words: organized water does not solubilize ions to the same extent that liquid water does. Therefore, the equilibrium concentration of these ions in the cellular water will be lower than in the free solution bathing the cells. This might be the case of

Na^+ ions. On the other hand, those ions that, because of the *c*-value of the fixed charges, are selectively adsorbed, may reach a higher concentration inside the cell than in the bathing solution. This might be the case of K^+ ions.

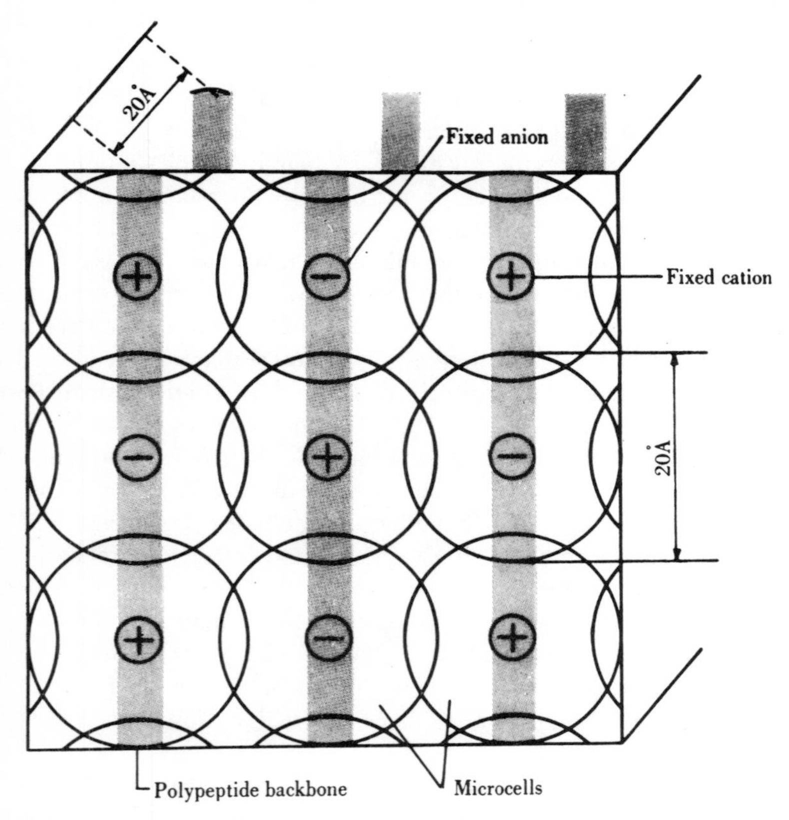

FIGURE 9–8 Hypothetical scheme of the fixed charge groups showing microcells of a radius of 10 Å. (Taken with kind permission from G.N. Ling, 1962.)

An analogous reasoning can be made for the H-binding of non-electrolytes, where the number, position of H-bonding sites and other stereochemical factors will dictate the selective accumulation of certain sugars, aminoacids, etc.

9.7 *The movement and distribution of substances associated to the metabolism*

Since according to the Association-Induction Hypothesis the asymmetrical distribution of substances on both sides of the cell membrane is not based on "uphill" movement but in the adsorption of solutes on specific sites (or exclusion of solutes from a non-liquid cell water), there is no need in spending metabolic energy in the translocation and accumulation (or exclusion) of substances. However, as discussed in detail in chapter 5, a large amount of experimental evidence demonstrates that the movement and distribution of substances *is* indeed associated to the metabolism of the cell. What is then the explanation of the Association-Induction Hypothesis for this fact?

The answer provided by the Association-Induction Hypothesis is based on the following points: a) A cardinal adsorbent in the right cardinal site can determine which (among several possible) *c*-value and affinities a molecule or a gang of associated sites will have. b) ATP, or other energy-rich intermediate, is needed. The metabolic machinery, or else the research worker if he has experimentally inhibited the metabolism, should afford the necessary ATP. c) However, on energetic grounds it can be asserted that it is not the *energy* stored in the phosphate linkages of ATP (or any other molecule that the metabolism might generate) what "pumps" the solutes. d) The cell has a state in which its ATP and K^+ content are high and its Na^+ content is low. This is the "healthy" state. When this is impaired (say by using a metabolic inhibitor), and the metabolic output does not match the ATP hydrolysis, the ATP content goes down and the cell loses K^+ and gains Na^+. Hence there is another cellular state in which ATP and K^+ are low and Na^+ high. But there is a long way out to say that it is the *energy* of the phosphate linkages what is being used to pump K^+ in, and Na^+ out of the cell.

The answer of the Association-Induction Hypothesis to why the accumulation and extrusion of certain molecular species are so closely dependent on the correct functioning of the metabolic machinery, is that each ATP molecule adsorbs on a cardinal site and this induces in a large number of sites (the gang) certain *c*-values that makes them selective for K^+ and other solutes. If the ATP content goes down (something like taking B^+ out of the system depicted in figures 9–4 to 9–6) the protoplasm will evolve to a non-energized state which no longer favours K^+ over Na^+ (or not to the "physiological" extent). In fact the Association-Induction Hypothesis postulates a delicate interaction between ATP production and ion content on the basis of the following scheme

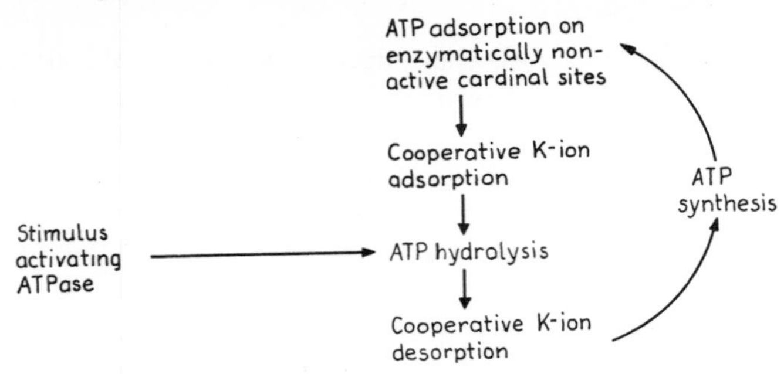

The Association-Induction Hypothesis does not maintain that active transport, as we have studied it in chapter 5 cannot occur at all. As we have emphasized in chapter 7, epithelial membranes show an unequivocal case of active transport from a truly free solution to another truly free solution. It is in the case of movement into an out of the single cells where the Hypothesis maintains that substances are not translocated from one free solution to the other and, therefore, there is no need for a pumping mechanism. Uptake phenomena like saturability and competition, that Membrane Theory explains on the basis of pores and carriers, are taken to reflect the existence of a limited number of fixed sites in the protoplasm.

The far reaching implications of this Hypothesis, in particular those concerning electrical phenomena, will not be discussed here. The derivation of equations for kinetic and steady state distributions as well as the experimental support of some of the ideas discussed can be consulted in the quoted papers of Ling.

General comments

BIOLOGICAL MEMBRANES are basic cellular structures with an elaborated molecular organization in dynamic equilibrium with the environment. As it becomes evident that enzymes systems attached to membranes possess properties that are lost upon purification, the participation of membranes in biological processes as fundamental as the synthesis of ATP, the selective accumulation of matter, the synthesis of proteins, the handling of nervous information or the perception of light, is viewed not as an incidental circumstance but as the very condition that makes them possible. Coupling between chemical reactions and fluxes of both, substances and energy at the membrane level, results in the improbable accumulation of matter that has always puzzled the students of the Second Law of Thermodynamics. Biology abstracts from the complexities of living systems and in their places substitutes somewhat idealized situations that are more amenable to analysis. In this sense a membrane working between two well stirred solutions, one of them being the protoplasm, has been a fruitful simplifying concept. By representing the profile of a cell membrane by a lipid bilayer and assuming it has some relatively simple translocating mechanisms (most of which are also known to exist in non-living systems) biophysicists have been so successful in affording plausible explanations for a large number of biological phenomena, that today cell membrane's profile ranks second only to Cleopatra's in fame, and we feel it will improve.

In describing membrane phenomena we have used the same assumptions and theoretical approaches used by the workers in the specific fields, but have also payed attention to scientists holding different positions. Thus we have described membranes separating well stirred compartments from chapter 2 to 8, but in chapter 9 we have described Ling's Association-Induction

Hypothesis which is one of the most fascinating models of modern biology, and which disregards the role of the membrane as a barrier. If, in order to achieve the typical distribution of intracellular substances, the membrane had to translocate solutes from a free solution to another, then the amount of energy required would exceed the output of energy of the cellular metabolism. This would had been a deadly critic to Membrane Theory had it not been followed by the demonstration that ions and water in the protoplasm are not entirely in free solution. In our view, if intracellular water is not entirely in free state, and ions are not completely dissociated, then much of the central criticism to Membrane Theory will not have a destructive effect. Even the extracellular space does not seem to contain in all the cases a truly free solution. So that the only thing that would vary in the membrane scope would be the magnitude of the electrochemical potential gradients across the membrane. Membrane Theory—as view by non-membranologists—considers the cell as a rubber balloon full of solution. Non-membrane Theories—as view by membranologists—consider the cell as a sponge inside a net. The present status of research suggests that truth might be somewhere between these extreme positions. Thus Kolber and Stein and Fox and Kennedy (see chapter 8) have in a test tube the pumps that non-membrane theories say that do not exist, and Ling (see chapter 9) took pictures of directional crystallization of the intracellular water that membranologists claim should be liquid.

As we said in the Introduction, we hope we have given the reader enough flexibility as to accept opinions and criticism from both sides.

Bibliography

Adrian, R.H., "Internal chloride concentration and chloride efflux of frog muscle", *J. Physiol.*, 1961, **156**, 623.

Adrian, R.H. and Freygang, W.H., The potassium and chloride conductance of frog muscle membrane", *J. Physiol.*, 1962(a), **163**, 61.

Adrian, R.H. and Freygang, W.H., "Potassium conductance of frog muscle membrane under controlled voltage", *J. Physiol.*, 1962(b), **163**, 104.

Agar, W.T., Hird, F.J.R. and Sidhu, G.S., "The active absorption of amino-acids by the intestine", *J. Physiol.*, 1953, **121**, 255.

Alvarado, F., "Transport of sugars and amino acids in the intestine: evidence for a common carrier", *Science*, 1966, **151**, 1010.

Ambrose, E.J., "Cell contacts", in *Rec. Prog. Surface Sci.* (ed. by Danielli, J.F., Pankhurst, K.G.A. and Riddiford, A.C.), Acad Press, New York, 1964.

Ambrose, E.J., James, A.M. and Lowick, J.A.B., "Differences between the electrical charge carried by normal and homologous tumour cells", *Nature*, 1956, **177**, 576.

Andersen, B., and Ussing, H.H., "Solvent drag on non-electrolytes during osmotic flow through isolated toad skin and its response to antidiuretic hormone", *Acta Physiol. Scand.*, 1957, **39**, 228.

Andreoli, T.E., Tieffenberg, M. and Tosteson, D.C., "The effect of valinomycin on the ionic permeability of thin lipid membranes", *J. Gen. Physiol.*, 1967, **50**, 2527.

Armstrong, W.McD. and Rothstein, A., "Discrimination between alkali metal cations by yeast. II. Cation interactions in transport", *J. Gen. Physiol.*, 1968, **50**, 967.

Askew, F.A. and Danielli, J.F., "Measurements of the pressures due to monolayers at oil-water interfaces", *Trans. Faraday Soc.*, 1940, **36**, 785.

Auditore, J.V., "Sodium-potassium activated G-strophanthin sensitive ATPase in cardiac muscle", *Proc. Soc. Exptl. Biol. Med.*, 1962, **110**, 595.

Aull, F. and Hempling, H.G., "Sodium fluxes in the Ehrlich mouse ascites tumor cell", *Am. J. Physiol.*, 1963, **204**, 789.

Aull, F. and Hempling, H.G., "Kinetic analysis of sodium fluxes in the Ehrlich mouse ascites tumor cell", in press, 1969.

Autilio, L.A., Norton, W.T., and Terry, R.D., "The preparation and some properties of purified myelin from the central nervous system", *J. Neurochem.*, 1964, **11**, 17.

Autilio, L.A., Norton, W.T., and Terry, R., quoted by Vandenheuvel (1965).

Baker, P.F., "Phosphorus metabolism of intact crab nerve and its relation to the active transport of ions", *J. Physiol.*, 1965, **180**, 383.

Baker, P.F., "Recent experiments on the properties of the Na efflux from squid axons", *J. Gen. Physiol.*, 1968, **51**, 172, suppl.

Baker, P.F., Hodgkin, A.L. and Shaw, T.I., "The effects of charges in internal ionic concentrations on the electrical properties of perfused giant axons", *J. Physiol.*, 1962, **164**, 355.

Bangham, A.D., de Gier, J. and Greville, G.D., "Osmotic properties and water permeability of phospholipid liquid crystals", *Chem. Phys. Lipids*, 1967, **1**, 225.

Bangham, A.D., Pethica, B.A. and Seaman, G.V.F., "Charged groups at the interface of some blood cells", *Biochem. J.*, 1958, **69**, 12.

Bangham, A.D., Standish, M.M. and Watkins, J.C., "Diffusion of univalent ions across the lamellae of swollen phospholipids", *J. Mol. Biol.*, 1965, **13**, 238.

Bangham, A.D., Standish, M.M., Watkins, J.C., and Weissmann, G., "The diffusion of ions from phospholipid model membrane systems", *Protoplasma*, 1967, **63**, 183.

Barrer, R.M., *Diffusion in and through solids*, MacMillan, New York, 1941.

Battaglia, F.C., and Randle, P.J., "Regulation of glucose uptake by muscle IV. The specificity of monosaccharide-transport systems in rat diaphragm muscle", *Biochem. J.*, 1960, **75**, 408.

Bear, R.S., Palmer, K.J. and Schmitt, F.O., "X-Ray diffraction studies of nerve lipids", *J. cell. comp. Physiol.*, 1941, **17**, 355.

Bennet, H. S., "The concept of membrane flow and membrane vesiculation as mechanisms for active transport and ion pumping", *J. Biophys. Biochem. Cytol.*, 1956, **2**, 99, suppl.

Berendsen, H. J. C., "Nuclear magnetic resonance study of collagen hydration", *J. Chem. Phys.*, 1962, **36**, 3297.

Berendsen, H. J. C., "Water structure", in *Theoretical and Experimental Biophysics.* (Ed. Cole, A.), M. Dekker, New York, 1967, Vol. I.

Berg, H. C., Diamond, J. M. and Marfey, P. S., "Erythrocyte membrane: chemical modification", *Science*, 1965, **150**, 64.

Bernal, J. D., "The structure of water and its biological implications", in *The state and movement of water in living organism.* Symposia Soc. Exper. Biol. XIX, (Ed. Fogg, G. E.), Cambridge, 1965.

Bernstein, J., *Electrobiologie*, Fr. Vieweg, Braunschweig (1912).

Biber, T. U. L., Chez, R. A. and Curran, P. F., "Na transport across frog skin at low external Na concentrations", *J. Gen. Physiol.*, 1966, **49**, 1161.

Bobinski, H., and Stein, W. D., "Isolation of a glucose-binding component from human erythrocyte membranes", *Nature*, 1966, **211**, 1366.

Bolingbroke, V., Harris, E. J. and Sjodin, R. A., "Rubidium and Caesium entry, and cation interaction in frog skeletal muscle", *J. Physiol.*, 1961, **157**, 289.

Bonting, S. L. and Caravaggio, L. L., "Studies on sodium-potassium-activated adenosinetriphosphatase. V. Correlation of enzyme activity with cation flux in six tissues", *Arch. Biochem. Biophys.*, 1963, **101**, 37.

Bourguet, J. and Morel, F., "Indépendance des variations de perméabilité à l'eau et au sodium produites par les hormones neurohypophysaires sur la vessie de grenouille", *Biochim. Biophys. Acta*, 1967, **135**, 693.

Bowyer, F. and Widdas, W. F., "The facilitated transfer of glucose and related compounds across the erythrocytes membrane", *Disc. Faraday Soc.* 1956, **21**, 251.

Bowyer, F. and Widdas, W. F., "The action of inhibitors on the facilitated hexose transfer system in erythrocytes", *J. Physiol.*, 1958, **141**, 219.

Boyer, P. D., Bieber, L. L., Mitchell, R. A. and Szabolcski, G., "Apparent independence of the phosphorylation and water-formation reactions from the oxidation reactions of oxidative phosphorylation", *J. Biol. Chem.*, 1966, **241**, 5384.

Brachet, J., *Biochemical Cytology*, Academic Press, N. York, 1957.

Brandt, P. W., "A study of the mechanism of pinocytosis", *Exptl. Cell. Res.*, 1958, **15**, 300.

Brandt, P.W. and Freeman, A.R., "The role of surface chemistry in the biology of pinocytosis", *J. Colloid Sci.*, 1967, **25**, 47.

Brandt, P.W. and Pappas, G.D., "An electronmicroscopic study of pinocytosis in ameba", *J. Biophys. Biochem. Cytol.*, 1960, **8**, 675.

Bregman, J.I., "Cation exchange processes", *Ann. New York Acad. Sci.*, 1954, **57**, 125.

Bricker, N.S., Biber, T. and Ussing, H., "Exposure of the isolated frog skin to high potassium concentrations at the internal surface. I) Bioelectric phenomena and sodium transport", *J. Clin. Invest.*, 1963, **42**, 88.

Brown, A.D. "Hydrogen ion titrations of intact and dissolved lipoprotein membranes", *J. Mol. Biol.*, 1965, **12**, 491.

Brown, H.D., Chattopadhyay, S.K. and Patel, A., "Sarcoplasmic reticulum ATPase on a solid support", *Biochem. Biophys. Res. Commun.*, 1966, **25**, 304.

Bueno, E.J., and Corchs, L., "Induced pacemaker activity on toad skin", *J. Gen. Physiol.*, 1968, **51**, 785.

Bungenberg de Jong, H.G., "Reversal of charge phenomena, equivalent weight and specific properties of ionized groups", *Colloid Science* **2** (Ed. Kruyt, H.R.,), Elsevier, Amsterdam, 1949.

Burger, M., Hejmová, L., and Kleinzeller, A., "Transport of some mono- and di-saccharides into yeast cells", *Biochem. J.*, 1959, **71**, 233.

Buvat, R., "Investigations on the infrastructures of the cytoplasm in the cells of the apical meristem of the leaf buds and of the developed leaves of elodea canadiensis", *Ann. Sci. Nat. Bot.*, 1958, **19**, 121.

Caldwell, P.C., "The effect of certain metabolic inhibitors on the phosphate esters of the squid giant axon", *J. Physiol.*, 1956, **132**, 35p.

Caldwell, P.C., "The phosphorus metabolism of squid axons and its relationship to the active transport of sodium", *J. Physiol.*, 1960, **152**, 545.

Caldwell, P.C., Hodgin, A.L. and Shaw, T.I., "Injection of compounds, containing energy-rich phosphate bond into giant nerve fibres", *J. Physiol.* 1959, **147**, 18p.

Caldwell, P.C., Hodgkin, A.L., Keynes, R.D. and Shaw, T.J., "The effects of injecting energy-rich phosphate compounds on the active transport of ions in the giant axons of loligo", *J. Physiol.*, 1960, **152**, 561.

Caldwell, P.C. and Keynes, R.D., "The effect of ouabain on the efflux of sodium from a squid giant axon", *J. Physiol.*, 1959, **148**, 8p.

Campbell, R.D., "Desmosome formation: an hypothesis of membrane accumulation", *Proc. Nat. Aacad. Sci.*, 1967, **58**, 1422.

Candia, O.A., "Active sodium transport in the isolated bulfrog cornea", *Biochim. Biophys. Acta*, 1968, **163**, 262.

Carr, Ch.W. and Sollner, K., "The electroosmotic effects arising from the interaction of the selectively anion and selectively cation permeable parts of mosaic membranes", *Biophysical J.*, 1964, **4**, 189.

Cass, A. and Finkelstein, A., "Water permeability of thin lipid membranes", *J. Gen. Physiol.* 1967, **50**, 1765.

Cerbón, I., "Nuclear magnetic resonance of water in microorganism", *Biochim. Biophys. Acta*, 1964, **88**, 444.

Cerbón, J., "Variations of the lipid phase of living mircoorganism during the transport process", *Biochim. Biophys. Acta*, 1965, **102**, 449.

Cerbón, J., "NMR studies on the water immobilization by lipid systems in vitro and in vivo", *Biochim. Biophys. Acta*, 1967, **144**, 1.

Cereijido, M. and Curran, P.F., "Intracellular electrical potentials in frog skin", *J. Gen. Physiol.*, 1965, **48**, 543.

Cereijido, M., Herrera, F.C., Flanigan, W.J., and Curran, P.F., "The influence of Na concentration on Na transport across frog skin", *J. Gen. Physiol.*, 1964, **47**, 879.

Cereijido, M., Migliora, G. and Rotunno, C.A., "Effect of antidiuretic hormone on the unidirectional flux of Na across the outer border of the epithelium", 1969 submitted.

Cereijido, M., Reisin, I. and Rotunno, C.A., "The effect of sodium concentration on the content and distribution of sodium in the frog skin", *J. Physiol.*, 1968, **196**, 237.

Cereijido, M. and Rotunno, C.A., "Transport and distribution of sodium across frog skin", *J. Physiol.*, 1967, **190**, 481.

Cereijido, M. and Rotunno, C.A., "Fluxes and distribution of sodium in frog skin: a new model", *J. Gen. Physiol.*, 1968, **51**, 280, suppl.

Cereijido, M., Rotunno, C.A. and Vilallonga, F.A., "Unidirectional flux of sodium across the outer border of the epithelium", 1969 submitted.

Cereijido, M., Vilallonga, F., Fernandez, M., and Rotunno, C.A., "On the interactions between lipids and ions", *Proc. Symp. Molecular basis of membrane function*, Duke, N. Carolina, 1968, Prentice Hall, 1969.

Chance, B., "Energy-linked cytochrome oxidation in mitochondria", *Nature*, 1961, **189**, 719.

Chance, B., Ghosh, A., Higgins, J. and Maitra, P., "Cyclic and oscillatory responses of metabolic pathways involving feedback and their computer representations", *Ann. N. Y. Acad. Sci.*, 1964, **115**, 1010.

Changeux, J.P., Thiéry, J., Tung, Y., and Kittel, C., "On the cooperativity of biological membranes", *Proc. Nat. Acad. Sci.*, 1967, **57**, 335.

Chapman, D., *The structure of lipids*, Methuen, London, 1965.

Chapman, D., "Liquid crystals and cell membranes", *Ann. N. York Acad. Sci.*, 1966, **137**, 745.

Chapman, G. and McLauglan, K.A., "Oriented water in the sciatic nerve of rabbit", *Nature*, 1967, **215**, 391.

Chapman-Andresen, C., quoted by Rustad, 1964.

Chapman-Andresen, C. and Holter, H., "The ingestion of carbin-14 glucose by pinocytosis in the amoeba chaos", *Exptl. Cell. Res. Suppl.*, 1955, **3**, 52.

Cheesman, D.F. and Davies, J.T., Physicochemical and biological aspects of proteins at interfaces", *Advan. Prot. Chem.*, 1954, **9**, 439.

Chez, A.R., Palmer, R.R., Schultz, S.G. and Curran, P.F., "Effect of inhibitors of alanine transport in isolated rabbit ileum", *J. Gen. Physiol.*, 1967, **50**, 2357.

Choi, J.K., "The fine structure of the urinary bladder of the toad Bufo marinus", *J. Cell. Biol.* 1963, **16**, 53.

Christensen, H.N., *Biological transport*, Benjamin, New York, 1962.

Christensen, H.N., and Riggs, T.R., Concentrative uptake of amino-acids by the Ehrlich mouse ascites carcinoma cell", *J. Biol. Chem.*, 1952, **194**, 57.

Ciani, S., quoted by Eisenman, Sandbloom and Walker, 1967.

Ciani, S. and Gliozzi, A., "Electrical properties of liquid ion-exchange membranes with dissociated sites", *Biophysik*, 1968, **5**, 145.

Cirillo, V.P., "The transport of non-fermentable sugars across the yeast cell membrane", in *Membrane Transport and Metabolism* (Eds. Kleinzeller, A. and Kotyk, A.) Acad. Press., London, 1960.

Civan, M.M. and Frazier, H.S., "The site of the stimulatory action of vasopressin on sodium transport in toad bladder", *J. Gen. Physiol.*, 1968, **51**, 589.

Clunie, J.S., Corkill, J.M., Goodman, J.F. and Ogden, C.P., "Conductance of foam films", *Trans. Faraday Soc.*, 1967, **63**, 505.

Cohen, D., "Specific binding of rubidium in chlorella", *J. Gen. Physiol.*, 1962, **45**, 959.

Cohen, G.N. and Monod, J., "Bacterial permeases", *Bacteriological Reviews*, 1957, **21**, 169.

Cohen, L.B., Keynes, R.D. and Hille, B., "Light scattering and birefringence changes during nerve activity", *Nature*, 1968, **218**, 438.

Cole, K.S., "Surface forces of the arbacia egg", *J. Cellular Comp. Physiol.*, 1932, **1**, 1.

Cole, K.S., "Dynamic electrical characteristics of the squid axon membrane", *Arch. Sci. Physiol.*, 1949, **3**, 253.

Cole, K.S., "Bioelectricity: electric physiology", in Glasser, O.: *Medical physics*, Vol.II. Year Book Publishers, Chicago, 1950.

Cole, K.S., "Electro diffusion models for the membrane of squid giant axon", *Physiol. Rev.*, 1965, **45**, 340.

Cole, K.S. and Curtis, H.J., "Electric impedance of the squid giant axon during activity", *J. Gen. Physiol.*, 1939, **22**, 649.

Collander, R., "The permeability of plant protoplasts to small molecules", *Physiol. Plantarum*, 1949, **2**, 300.

Conti, F. and Eisenman, G., The non-steady state membrane potential of ion exchangers with fixed sites", *Biophys. J.*, 1965a, **5**, 247.

Conti, F. and Eisenman, G., "The steady state properties of ion exchange membranes with fixed sites", *Biophys. J.*, 1965b, **5**, 511.

Conti, F. and Eisenman, G., "The steady-state properties of an ion exchange membrane with mobile sites", *Biophys. J.*, 1966, **6**, 227.

Conway, E.J., "Ionic permeability of skeletal muscle fibres", *Nature*, 1946, **157**, 715.

Conway, E.J. and Duggan, F., "A cation carrier in the yeast cell wall", *Biochem. J.*, 1958, **69**, 265.

Conway, E.J. and Cruess-Callaghan, G., "Magnesium and Chloride 'permeations' in muscle", *Biochem. J.*, 1937, **31**, 828.

Cook, G.M.H., Heard, D.H. and Seaman, G.V.F., "The electrokinetic, characterization of the Ehrlich ascites carcinoma cell", *Exptl. Cell. Res.* 1962, **28**, 27.

Cope, F.W., "Solid state mechanisms of electron and ion transport in biological systems", *Symposium on electrochemical aspects of molecular biology*, 1965.

Cope, F.W., "Evidence for complexing of Na^+ in muscle, kidney and brain, and by actomyosin. The relation of cellular complexing of Na^+ to water structure and to transport kinetics", *J. Gen. Physiol.*, 1967, **50**, 1353.

Cornwell, D.G., Heikkila, R.E., Bar, R.S. and Biagi, G.L., "Red blood cell lipids and the plasma membrane", *J. Am. Oil. Chem. Soc.*, 1967, **45**, 5.

Cowan, S.L., "The action of potassium and other ions on the injury potential and action current in maia nerve", *Roy. Soc. Proc. Series B. Biol. Sci.* 1934, **115**, 216.

Cowie, D.B. and Roberts, R.B., "Permeability of micro-organisms to inorganic ions, amino acids and peptides", in *Electrolytes in biological systems* (ed. Shanes, A.M.), Am. Physiol. Soc. Washington, D.C. 1955, 1.

Crabbé, J. and DeWeer, P., "Action of aldosterone and vasopressin on the active transport of sodium by the isolated toad bladder", *J. Physiol.*, 1965, **180**, 560.

Crane, R.K., "Intestinal absorption of sugars", *Physiol. Rev.*, 1960, **40**, 789.

Crane, R.K., "Hypothesis for mechanism of intestinal active transport of sugars", *Fed. Proc.*, 1962, **21**, 891.

Crane, R.K., "Na-dependent transport in the intestine and other animal tissues", *Fed. Proc.*, 1965, **24**, 1000.

Crane, R.K., "Structural and functional organization of an epithelial cell brush border", in *Intracellular transport* (Ed. Warren, K.B.), Academic Press, New York, 1966.

Crane, R.K., and Mandelstam, P., "Active transport of sugar and metabolism in different parts of hamster intestine", *Fed. Proc.*, 1959, **18**, 208.

Crane, R.K., Miller, D. and Bihler, I., "The restrictions on possible mechanisms of intestinal active transport of sugars", *Symposium on membrane transport and metabolism* (Eds. Keinzeller, A. and Kotyk, A.), 1961.

Csáky, T.Z., "Significance of sodium ions in active intestinal transport of nonelectrolytes", *Am. J. Physiol.*, 1961, **201**, 999.

Csáky, T.Z. and Zollicoffer, L., "Ionic effect on intestinal transport of glucose in the rat", *Am. J. Physiol.*, 1960, **198**, 1056.

Curie, P., *Œuvres*, Gauthier-Villars, Paris, 1908.

Curran, P.F., "Na, Cl, and water transport by rat ileum in vitro", *J. Gen. Physiol.*, 1960, **43**, 1137.

Curran, P.F., "Ion transport in intestine and its coupling to other transport processes", *Fed. Proc.*, 1965, **24**, 993.

Curran, P.F. and Cereijido, M., "K fluxes in frog skin", *J. Gen. Physiol.*, 1965, **48**, 1011.

Curran, P.F., Herrera, F.C., and Flanigan, W.S., "The effect of Ca and antidiuretic hormone on Na transport across frog skin. II. Sites and mechanisms of action", *J. Gen. Physiol.*, 1963, **46**, 1011.

Curran, P.F. and MacIntosh, J.R., "A model system for biological water transport", *Nature*, 1962, **193**, 347.

Curran, P.F., Schultz, S.G., Chez, R.A. and Fuisz, R.E., "Kinetic relations of the Na-amino acid interaction at the mucosal border of intestine", *J. Gen. Physiol.*, 1967, **50**, 1261.

Curran, P.F. and Solomon, A.K., "Ion and water fluxes in the ileum of rats", *J. Gen. Physiol.*, 1957, **41**, 143.

Dainty, J. and House, C.R., "Unstirred layers in frog skin", *J. Physiol.*, 1966, **182**, 66.

Dainty, J. and Ginzburg, B.Z., "Irreversible thermodynamics and frictional models of membrane processes, with particular reference to the cell membrane", *J. Theoret. Biol.*, 1963, **5**, 256.

Danford, M.D. and Levy, H.A., "The structure of water at room temperature", *J. Am. Chem. Soc.*, 1962, **84**, 3965.

Danielli, J.F., "Morphological and molecular aspects of active transport", *Symp. Soc. Exp.*, 1954, **8**, 502.

Danielli, J.F. and Harvey, E.N., "The tension at the interface of mackerel egg oil, with remarks on the nature of the cell surface", *J. Cellular Comp. Physiol.*, 1935, **5**, 483.

Davies, J.T., "Interfacial potentials. II. Molecular orientations of substituted fatty acids and of polyamino acids", *Trans. Faraday Soc.*, 1953, **49**, 949.

Davies, J.T. and Rideal, E.K., *Interfacial phenomena*, Academic Press, 1961, New York.

Davson, H. and Danielli, J.F., *The permeability of natural membranes*, Cambridge, 1943.

Davson, H., and Reiner, J.M., "Ionic permeability: an enzyme-like factor concerned in the migration of sodium through the cat erythrocyte membrane", *J. Cell. Com. Physiol.*, 1942, **20**, 325.

Dawson, A.C. and Widdas, W.F., "Inhibition of glucose permeability of human erythrocytes by N-etaylmaleimide", *J. Physiol.*, 1963, **168**, 644.

Dawson, A.C. and Widdas, W.F., "Variations with temperature and pH of the parameters of glucose transfer across the erythrocyte membrane in the foetal guinea-pig", *J. Physiol.*, 1964, **172**, 107.

Dawson, R.M.C., Hemington, N. and Lindsay, D.B., "The phospholipids of the erythrocyte ghosts of various species", *Biochem. J.*, 1960, **77**, 226.

Deamer, D.W. and Packer, L., "Correlation of ultrastructure with light-induced ion transport in chloroplasts", *Arch. Biochem. Biophys.*, 1967, **119**, 83.

Deamer, D.W., Utsumi, K. and Packer, L., "Oscillatory states of mitochondria. III. Ultrastructure of trapped conformational states", *Arch. Biochem. Biophys.*, 1967, **121**, 641.

De Busk, B.G. and De Busk, A.G., "Molecular transport in neurospora crassa. I. Biochemical properties of phenylalanine permease", *Biochim. Biophys. Acta*, 1965, **104**, 139.

15 Cereijido/Rotunno (0241)

de Donder, T. and van Rysselberghe, P., *Thermodynamic theory of affinity*, Stanford University Press, Stanford, 1936.

de Gier, J. and van Deenen, L.L.M., "Some lipid characteristics of red cell membranes of different species", *Biochim. Biophys. Acta*, 1961, **49**, 286.

de Groot, S.R., *Thermodynamics of irreversible processes*, North-Holland, Amsterdam, 1951.

Dekker, A.J., *Solid state physics*, Englewood cliffs, Prentice-Hall, New Jersey, 1957.

Denbigh, K.G., *The thermodynamics of the steady state*, Methuen, London, 1951.

De Robertis, E., "Ultrastructure and cytochemistry of the synaptic region", *Science*, 1967, **156**, 3777.

De Robertis, E., Nowinsky, W. and Saez, F., *Cell Biology*, Saunders, Philadelphia, 1969.

Dervichian, D.G., "The structure of the protein molecule", *J. Chem. Phys.*, 1943, **11**, 236.

Dervichian, D.G., "Swelling and molecular organization in colloidal electrolytes", *Trans. Faraday Soc.*, 1946, **42B**, 180.

Dervichian, D.G., "The existence and significance of molecular associations in monolayers", in *Surface Phenomena in Chemistry and Biology* (Eds. Danielli, J.F., Pankhurst, K.G.H. and Riddiforf, A.C.), Pergamon, London, 1958.

Dervichian, D.G., "The physical chemistry of phospholipids", in *Progress in Biophysics and Molecular Biology* (Eds. Butler, J.A.V. and Huxley, H.E.) Pergamon Press, London, 1964. Vol. 14.

De Terra, N. and Rustad, R.C., "The dependence of pinocytosis on temperature and aerobic respiration", *Exptl. Cell Res.*, 1959, **17**, 191.

De Vries, H., "Eine Methode zur Analyse der Turgorkraft", *Jahrb. Wiss. Botan.*, 1884, **14**, 427.

De Vries, H., "Plasmolytische Studien über die Wand der Vakuolen", *Jahrb. Wiss. Botan.*, 1885, **16**, 465.

Diamond, J.M., "The mechanism of solute transport by the gall bladder", *J. Physiol.*, 1962, **161**, 474.

Diamond, J.M., "Transport of salt and water in rabbit and guinea pig gall bladder", *J. Gen. Physiol.*, 1964, **48**, 1.

Diamond, J.M. and Bossert, W.M., "Standing-gradient osmotic flow", *J. Gen. Physiol.*, 1967, **50**, 2061.

Diamond, J. M. and Solomon, A. K., "Intracellular potassium compartments in nitella axilaris", *J. Gen. Physiol.*, 1959, **42**, 1105.

Diamond, J. M. and Tormey, J. Mc. D., "Studies on the structural basis of water transport across epithelial membranes", *Fed. Proc.*, 1966, **25**, 1458.

Diamond, J. M. and Wright, E., *Ann. Rev. Physiol.*, 1969 (in press).

Dickman, S. R. and Speyer, J.F., "Factors affecting the activity of mitochondrial and soluble aconitase", *J. Biol. Chem.*, 1954, **206**, 67.

Dodge, J.T. and Philipps, G.B., "Composition of phospholipids and phospholipid fatty acids and aldehydes in human red cells", *J. Lipid. Res.*, 1967, **8**, 667.

Dourmashkin, R. R., Dougherty, R.M. and Harris, R.J. C., "The effect of saponin and digitonin on Rous sarcoma virus and cell membranes: an approach to the cytochemistry of cellular components by the use of negative staining", *J. Roy. Microsc. Soc.*, 1962, **81**, 215.

Dujardin (1835), quoted by Ling (1962).

Dunham, E. T., "Linkage of active cations transport to ATP utilization", *Physiologist*, 1957, **1**, 23.

Dunham, E. T. and Glynn, I. M., "Adenosinetriphosphatase activity and the active movements of alkali metal ions", *J. Physiol.*, 1961, **156**, 274.

Durbin, R. P., "Osmotic flow of water across permeable cellulose membranes", *J. Gen. Physiol.*, 1960, **44**, 315.

Durbin, R. P., Frank, H. and Solomon, A. K., "Water flow through frog gastric mucose", *J. Gen. Physiol.*, 1956, **39**, 535.

Durbin, R. P. and Heinz, E., "Electromotive chloride transport and gastric acid secretion in the frog", *J. Gen. Physiol.*, 1958, **41**, 1035.

Durbin, R. P. and Moody, F., "Water movement through a transporting epithelial membrane, the gastric mucosa", *Symp. Soc. Exptl. Biol.*, 1965, **19**, 299.

Edelman, I. S., Bogoroch, R. and Porter, G. A., "On the mechanism of action of aldosterone on sodium transport: the role of protein synthesis", *Proc. Nat. Acad. Sc.*, 1963, **50**, 1169.

Einstein, A., "Über die von der molekularkinetischen Theorie der Wärme geforderte Bewegung von in ruhenden Flüssigkeiten suspendierten Teilchen", *Ann. Physik.*, 1905, **17**, 549.

Eisenman, G., "On the elementary atomic origin of equilibrium ionic specificity", in *Membrane transport and metabolism* (eds. Kleinzeller, A. and Kotyk, A.), Academic Press, New York, 1961.

Eisenman, G., "Cation selective glass electrodes and their mode of operation", *Biophys. J.*, 1962, **2**, 259.

Eisenman, G., *Glass electrodes for hydrogen and other cations*, M. Dekker, New York, 1967a.

Eisenman, G. (1967b), quoted by Eisenman, Sandblom and Walker (1967).

Eisenman, G., Bates, R., Mattock, G. and Friedman, S.M., *The glass electrode*, J. Wiley, New York, 1965.

Eisenman, G., Ciani, S.M. and Szabo, G., "Some theoretically expected and experimentally observed properties of lipid bilayer membranes containing neutral molecular carriers of ions", *Fed. Proc.*, 1968, **27**, 1289.

Eisenman, G. and Conti, F., "Some inplications for biology of recent theoretical and experimental studies on ion permeation in model membranes", *J. Gen. Physiol.*, 1965, **48**, 65 (part 2).

Eisenman, G., Rudin, D.O. and Casby, J.U., "Glass electrode for measuring sodium ion", *Science*, 1957, **126**, 831.

Eisenman, G., Sandblom, J.P. and Walker, J.L., "Membrane structure and ion permeation", *Science*, 1967, **155**, 965.

Elbers, P.F., "The cell membrane: image and interpretation", in *Recent Progress in Surface Science* (Eds. Danielli, J.F., Pankhurst, K.G.A. and Riddiford, A.C.), Academic Press, New York, 1964.

Elul, R., "Fixed charge in the cell membrane", *J. Physiol.*, 1967, **189**, 351.

Emmelot, P., and Bos, C.J., "Adenosine triphosphatase in the cell-membrane fraction from the rat liver", *Biochim. Biophys. Acta*, 1962, **58**, 373.

Emmelot, P., Bos, C.J., Benedetti, E.L. and Rumke, P., "Studies on plasma membranes. I. Chemical composition and enzyme content of plasma membranes isolated from rat liver", *Biochim. Biophys. Acta*, 1964, **90**, 126.

Engbaek, L. and Hoshiko, T., "Electrical potential gradients through frog skin", *Acta Physiol. Scand.*, 1957, **39**, 348.

Essig, A., Frazier, H.S. and Leaf, A., "Evidence for 'electrogenic' active sodium in an epithelial membrane", *Nature*, 1963, **197**, 701.

Essig, A., Kedem, O and Hill, T.L., "Net flow and tracer flow in lattice and carrier models", *J. Theoret. Biol.*, 1966, **13**, 72.

Eylar, E.H., Madoff, M.A., Brody, O.V. and Oncley, J.L., "The contribution of sialic acid to the surface charge of the erythrocyte", *J. Biol. Chem.*, 1962, **237**, 1992.

Farquhar, M.G. and Palade, G.E., "Junctional complexes in various epithelia", *J. Cell. Biol.*, 1963, **17**, 375.
Farquhar, M.G. and Palade, G.E., "Cell junction in amphibian skin", *J. Cell. Biol.*, 1965, **26**, 263.
Farquhar, M.G. and Palade, G.E., "Adenosine triphosphatase in amphibian epidermis", *J. Cell. Biol.*, 1966, **30**, 359.
Fawcett, D.W., "Physiologically significant specializations of the cell surface", *Circulation*, 1962, **26**, 1105.
Fergason, J.L. and Brown, G.H., "Liquid crystals and living Systems", *J. Am. Oil Chem. Soc.*, 1967, **45**, 120.
Fernández Morán, H., *Fine structure of biological lamellar systems, in Biophysical Science* (Ed. Oncley, J.L.) Wiley, New York, 1959.
Fernández Morán, H., "Cell-membrane ultrastructure. Low-temperature electron microscope and X-ray diffraction studies of lipoprotein components in lamellar systems", *Circulation*, 1962, **26**, 1039.
Fernández Morán, H. and Finean, J.B., "Electron microscope and low-angle X-ray diffraction studies of the nerve myelin sheath", *J. Biophys. Biochem. Cytol.*, 1957, **3**, 725.
Finean, J.B., "X-ray diffraction studies on the polymorphism of phospholipids", *Biochim. Biophys. Acta*, 1953, **10**, 371.
Finean, J.B., "The role of water in the structure of peripheral nerve myelin", *J. Biophys. Biochem. Cytol.*, 1957, **3**, 95.
Finean, J.B., "Electron microscope and X-ray diffraction studies of the effects of dehydration on the structure of nerve myelin", *J. Biophys. Biochem. Cytol.*, 1960, **8**, 13.
Finean, J.B., "The nature and stability of the plasma membrane", *Circulation*, 1962, **26**, 1151.
Finean, J.B. and Rumsby, M.G., "Negatively stained lipoprotein membranes: the validity of the images", *Nature*, 1963, **197**, 1326.
Finkelstein, A., "Electrical excitability of isolated frog skin and toad bladder", *J. Gen. Physiol.*, 1964, **47**, 545.
Finkelstein, A. and Cass, A., "Permeability and electrical properties of thin lipid membranes", *J. Gen. Physiol.*, 1968, **52**, 145.
Finkelstein, A. and Mauro, M., "Equivalent circuits as related to ionic systems", *Biophysical J.*, 1963, **3**, 215.
Finkelstein, J.D. and Schachter, D., "Active transport of calcium by intestine: effects of hypophysectomy and growth hormone", *Amer. J. Physiol.*, 1962, **203**, 873.

Fischer, S., Cellino, M., Gariglio, P. and Tellez-Nagel, I., "Protein and RNA metabolism of squid axon (Dosidicus gigas)", *J. Gen. Physiol.*, 1968, **51**, 72, suppl.

Fogg, G.E. (Ed.), "The state and movement of water in living organisms", *Symp. Soc. Exper. Biol.*, Cambridge, 1965. Vol. 19.

Fowler, R. and Guggenheim, E.A., *Statistical Thermodynamics*, Cambridge, 1956.

Fox, C.F. and Kennedy, E.P., "Specific labeling and partial purification of the M protein, a component of the β-galactoside transport system of Escherichia coli", *Proc. of Natl. Acad. of Sciences*, 1965, **54**, 891.

Frank, F.C., "Liquid crystals: theory of liquid crystals", *Disc. Faraday Soc.*, 1958, **25**, 19.

Frank, H.S., "The structure of water", *Fed. Proc.*, 1965, **24**, S 1.

Frank, H.S. and Wen, W.Y., "Ion-solvent interaction: structural aspects of ion-solvent interaction in aqueous solutions; a suggested picture of water structure", *Disc. Faraday Soc.*, 1957, **24**, 133.

Franck, J., and Meyer, J.E., "An osmotic diffusion pump", *Arch. Biochem.*, 1947, **14**, 297.

Frazier, H.S., "The electrical potential profile of the isolated toad bladder", *J. Gen. Physiol.*, 1962, **45**, 515.

Frazier, H.S., Dempsey, E.F. and Leaf, A., "Movement of sodium across the mucosal surface of the isolated toad bladder and its modification by vasopressin", *J. Gen. Physiol.*, 1962, **45**, 529.

Frazier, H.S. and Leaf, A., "The electrical characteristics of active sodium transport in the toad bladder", *J. Gen. Physiol.*, 1963, **46**, 491.

Frazier, H.S. and Leaf, A., "Cellular mechanisms in the control of body fluids", *Medicine (Baltimore)*, 1964, **43**, 281.

Frenkel, J.I., *Kinetic theory of liquids*, Oxford Univ. Press, London, 1946.

Freygang, W.H., "Tubular ionic movements", *Fed. Proc.*, 1965, **24**, 1135.

Freygang, W.H., Goldstein, D.A. and Hellam, D.C., "The after-potential that follows trains of impulses in frog muscle fibers", *J. Gen. Physiol.*, 1964, **47**, 929.

Frey-Wyssling, A., *Submicroscopic morphology of protoplasm*, Elsevier, Amsterdam, 1953.

Fricke, H., "The electric capacity of suspensions with special references to blood", *J. Gen. Physiol.*, 1925, **9**, 137.

Fricke, H., "Electric impedance of suspensions of biological cells", *Cold Spring Harbor Symposia*, 1933, **1**, 117.

Fridhandler, L. and Quastel, J. H., "Absorption of amino acids from isolated surviving intestine", *Arch. Biochem. Biophys.*, 1955, **56**, 424.

Friedenberg, R. M., *The electrostatics of biological cell membranes*, North-Holland, Amsterdam, 1967.

Fritz, O. G. and Swift, T. J., "The state of water in polarized and depolarized frog nerves. A proton magnetic resonance study", *Biophys. J.*, 1967, **7**, 675.

Frumento, A. S., "Sodium pump: its electrical effects on skeletal muscle", *Science*, 1965, **147**, 1442.

Fukuda, T. R., "Über die Bedingungen für das Zustandekommen des Asymmetriepotentials der Froschhaut", *Jap. J. Med. Sci.*, Part 3, 1942, **8**, 123.

Fukui, S., and Hochster, R. M., "Carbohydrate inhibitors of sucrose uptake by resting cells of Agrotacterium tumefaciens", *Canadian J. of Biochem.*, 1964, **42**, 1023.

Fukuta, N. and Mason, B. J., "Epitaxial growth of ice on organic crystals", *Phys. Chem. Solids*, 1963, **24**, 715.

Furchgott, R. F. and Ponder, E., "Electrophoretic studies in human red blood cells", *J. Gen. Physiol.*, 1941, **24**, 447.

Gale, E. F., "Assimilation of amino-acids by gram-positive bacteria and some actions of antibiotics thereon", *Advances in Protein Chemistry*, 1953, **8**, 287.

Galeotti, G., "Concerning the EMF which is generated at the surface of animal membranes on contact with different electrolytes", *Z. Physic. Chem.*, 1904, **49**, 542.

Galeotti, G., "Ricerche di elettrofisiologia secondo i criteri dell'elettrochimica", *Z. Allg. Physiol.*, 1907, **6**, 99.

Gardos, G., "Accumulation of K ions in human blood cells", *Acta Physiol. Acad. Sci. Hung.*, 1954, **6**, 191.

Garrahan, P. J. and Glynn, I. M., "The behaviour of the sodium pump in red cells in the absence of external potassium", *J. Physiol.*, 1967a, **192**, 159.

Garrahan, P. J. and Glynn, I. M., "The sensitivity of the sodium pump to external sodium", *J. Physiol.*, 1967b, **192**, 175.

Garrahan, P. J., and Glynn, I. M., "Factors affecting the relative magnitudes of the Na:K and Na:Na exchanges catalysed by the sodium pump", *J. Physiol.*, 1967c, **192**, 189.

Garrahan, P. J. and Glynn, I. M., "The incorporation of inorganic phosphate into adenosinetriphosphate by reversal of the sodium pump", *J. Physiol.*, 1967d, **192**, 237.

Geren, B.B., "The formation from the Schwann cell surface of myelin in peripheral nerves of chick embryos", *Exptl. Cell. Research.*, 1954, **7**, 558.

Gey, G.O., Shapras, P. and Borysko, E., "Activities and responses of living cells and their components as recorded by cinephase microscopy and electron microscopy", *Ann. N. Y. Acad. Sci.*, 1954, **58**, 1089.

Giebel, O., and Passow, H., "Die Permeabilität der Erythrocytenmembran für organische Anionen; zur Frage der Diffusien durch Poren", *Arch. Ges. Physiol.*, 1960, **271**, 378.

Giebisch, G., "Some electrical properties of single renal tubule cells", *J. Gen. Physiol.*, 1968, **51**, 315, suppl.

Ginzburg, B.Z. and Hogg, J., "What does a short circuit current measure in biological systems?", *J. Theoret. Biol.*, 1967, **14**, 316.

Glasstone, S., Laidler, K.J. and Eyring, H., *The theory of rate processes*, McGraw-Hill, N.Y. 1941.

Glauert, A.M., Dingle, J.T. and Lucy, J.A., "Action of saponin on biological cell membranes", *Nature*, 1962, **196**, 953.

Glauert, A.M. and Lucy, J.A., "Globular Micelles and the organization of membrane lipids", in *The Membranes* (Eds. Dalton, A.J. and Hagenan, F.) Academic Press, N.York, 1968.

Glynn, I.M., "The action of cardiac glycosides on sodium and potassium movements in human red cells", *J. Physiol.*, 1957, **136**, 148.

Glynn, I.M., "Activation of adenosinetriphosphatase activity in a cell membrane by external potassium and internal sodium", *J. Physiol.*, 1962a, **160**, 18 P.

Glynn, I.M., "An adenosine triphosphatase from electric organ activated by sodium and potassium and inhibited by ouabain or oligomycin", *Biochem. J.*, 1962b, **84**, 75 P.

Glynn, I.M., "Membrane adenosine triphosphatase and cation transport", *British Medical Bulletin*, 1968, **24**, 165.

Goldman, D.E., "Potential impedance, and rectification in membranes", *J. Gen. Physiol.*, 1943, **27**, 37.

Goldman, D.E., "A molecular structural basis for the excitation properties of axons", *Biophys. J.*, 1964, **4**, 189.

Goldman, R., Silman, H.I., Caplan, S.R., Kedem, O. and Katchalsky, E., "Papain membrane on a collodion matrix: preparation and enzymic behaviour", *Science*, 1965, **150**, 758.

Goldstein, D.A. and Solomon, A.K., "Determination of equivalent pore radius for human red cells by osmotic pressure measurement", *J. Gen. Physiol.*, 1960, **44**, 11.

Gonzalez, C.F., Shamoo, Y.E., Wyssbrod, H.R., Solinger, R.E. and Brodsky, W.A., "Electrical nature of sodium transport across the isolated turtle bladder", *Amer. J. Physiol.*, 1967, **213**, 333.

Gorter, E. and Grendel, F., "On bimolecular layers of lipoids on the chromocytes of the blood", *J. Exptl. Med.*, 1925, **41**, 439.

Gosselin, R.E., " The uptake of radiocolloids by macrophages in vitro. A kinetic analysis with radioactive colloidal gold", *J. Gen. Physiol.*, 1956, **39**, 625.

Green, D.E. and Fleischer, S., "The role of lipids in mitochondrial electron transfer and oxidative phosphorylation", *Biochim. Biophys. Acta.*, 1963, **70**, 554.

Green, K., "Ion transport across the isolated rabbit cornea", *Exp. Eye Res.*, 1966, **5**, 106.

Gregor, H.P. and Bregman, J.I., "Studies on ion exchange resins. IV. Selectively coefficients of various cation exchangers towards univalent cations", *J. Colloid. Sci.*, 1951, **6**, 323.

Guggenheim, E.A., *Thermodynamics*, Amsterdam, North Holland, 1950.

Gulik-Krzywicki, T., Rivas, E. et Luzzati, V., "Structure et polimorphisme des lipides: études par diffraction des rayons X du système formé de lipides de mitochondries de cœur de boeuf et d'eau", *J. Molec. Biol.*, 1967, **27**, 303.

Hackenbrock, C.R., "Ultrastructural basis for metabolically linked mechanical activity in mitochondria. I. Reversible ultrastructural changes with change in metabolic steady state in isolated liver mitochondria", *J. Cell. Biol.*, 1966, **30**, 269.

Hammett, L.P., *Physical organic chemistry*, McGraw-Hill, New York, 1940.

Hanai, T., Haydon, D.A. and Redwood, W.R., "The water permeability of artificial bimolecular leaflets: a comparison of radio-tracer and osmotic methods", *Ann. N. York. Acad. Science*, 1966, **137**, 731.

Hanai, T., Haydon, D.A. and Taylor, J., "Some further experiments on bimolecular lipid membranes", *J. Gen. Physiol.*, 1965, **48**, S 59.

Harkins, W.D., *The physical chemistry of surfaces*, Reihold, New York, 1952.

Harris, E.J., *Transport and accumulation in biological systems*, Butterworths Publications, London, 1960.

Harris, E.J. and Burns, G.P., "The transfer of sodium and potassium ions between muscle and the surrounding medium", *Trans. Faraday Soc.*, 1949, **45**, 508.

Harris, R.A., Penniston, J.T., Asai, J. and Green, D.E., "The conformational basis of energy conservation in membrane systems. II. Correlation between conformational change and functional states", *Proc. Nat. Acad. Sci.*, 1968, **59**, 830.

Harvey, E.N., "The tension at the surface of marine eggs especially those of the sea urchin", *Bull. Marine Biol. Lab.*, 1931, **61**, 273.

Harvey, E.N. and Danielli, J.F., "Properties of the cell surface", *Biol. Rev. Cambridge Phil. Soc.*, 1938, **13**, 319.

Hasselbach, W., "The calcium pump in membranes of the sarcoplasmic reticulum", *Proc. of the International Union of Physiol. Sciences; XXIV Int. Congress, 1968*, VI, 31.

Haydon, D.A., "The surface charge of cells and some other small particles as indicated by electrophoresis. I. The zeta potential-surface charge relationships", *Biochim. Biophys. Acta*, 1961, **50**, 450.

Haydon, D.A. and Taylor, J., "The stability and properties of bimolecular lipid leaflets in aqueous solutions", *J. Theoret. Biol.*, 1963, **4**, 281.

Hays, R.M. and Leaf, A., "Studies on the movement of water through the isolated toad bladder and its modification by vasopressin", *J. Gen. Physiol.* 1962a, **45**, 905.

Hays, R.M. and Leaf, A., "The state of water in the isolated toad bladder in the presence and absence of vasopressin", *J. Gen. Physiol.*, 1962b, **45**, 933.

Heard, D.H. and Seaman, G.V.F., "The influence of pH and ionic strength on the electro-kinetic stability of the human erythrocyte membrane", *J. Gen. Physiol.*, 1960, **43**, 635.

Hearst, J.E. and Vinograd, J., "The net hydration of T-4 bacteriophage deoxyribonucleic acid and the effect of hydration on buoyant behaviour in a density gradient at equilibrium in the ultracentrifuge", *Proc. Natl. Acad. Sci.*, 1961, **47**, 1005.

Hechter, O., "Intracellular water structure and mechanisms of cellular transport", *Ann. N.Y. Acad. Sci.*, 1965a, **125**, 625.

Hechter, O., "Role of water structure in the molecular organization of cell membranes", *Fed. Proc.*, 1965b, **24**, S-91.

Hechter, O. and Lester, G., "Cell permeability and hormone action", in *Recent Progress in Hormone Research* (Eds. Pincus, G.) Academic Press, New York, 1960.

Hedin, S. G., "Über die Permeabilität der Blutkörperchen", *Arch. Ges. Physiol.*, 1897, **68**, 229.

Hedin, S. G., "Versuche über das Vermögen der Salze einiger Stickstoffbasen, in die Blutkörperchen einzudringen", *Arch. Ges. Physiol.*, 1898, **70**, 525.

Heinz, E. and Durbin, R., "Evidence for an independent hydrogen-ion pump in the stomach", *Biochim. Biophys. Acta*, 1959, **31**, 246.

Helfferich, F., "Bi-ionic potentials", *Disc. Faraday Soc.*, 1956, **21**, 83.

Helfferich, F., *Ion Exchange*, McGraw-Hill, New York, 1962.

Henn, F. A., Decker, G. L., Greenawalt, J. W. and Thompson, T. E., "Properties of lipid bilayer membranes separating two aqueous phases: electron microscope studies", *J. Mol. Biol.*, 1967, **24**, 51.

Herrera, F. C., "Effect of insulin on short circuit current and sodium transport across toad urinary bladder", *Am. J. Physiol.*, 1965, **209**, 819.

Hertzenberg, L. A. and Hertzenberg, L. A., "Association of H-2 antigens with the cell membrane fraction of mouse liver", *Proc. Natl. Acad. Sci.*, 1961, **47**, 762.

Hill, A. V., *Chemical wave transmission in nerve*, Cambridge Univ. Press, London, 1932.

Hinke, J. A. M., "Glass microelectrodes for measuring intracellular activities of sodium and potassium", *Nature*, 1959, **184**, 1257.

Hladky, S. B., "The single file model for the diffusion of ions through a membrane", *Bull. Math. Biophys.*, 1965, **27**, 79.

Hladky, S. B. and Harris, J. D., "An ion displacement membrane model", *Biophysics J.*, 1967, **7**, 535.

Höber, R., *Physical chemistry of cells and tissues*, Churchill, London, 1945.

Hodgkin, A. L., "The effect of potassium on the surface membrane of an isolated axon", *J. Physiol.*, 1947, **106**, 319.

Hodgkin, A. L., "Ionic movements and electrical activity in giant nerve fibres", *Proc. Roy. Soc. B.*, 1958, **148**, 1.

Hodgkin, A. L. and Horowicz, P., "Movements of Na^+ and K^+ in single muscle fibres", *J. Physiol.*, 1959, **145**, 432.

Hodgkin, A. L. and Huxley, A. F., "Action potentials recorded from inside a nerve fibre", *Nature*, 1939, **144**, 710.

Hodgkin, A. L. and Huxley, A. F., "Currents carried by sodium and potassium ions through the membrane of the giant axon of Loligo", *J. Physiol.*, 1952, **116**, 449.

Hodgkin, A. L., Huxley, A. F. and Katz, B., "Ionic currents underlying activity in the giant axon of the squid", *Arch. Sci. Physiol.*, 1949, **3**, 129.

Hodgkin, A.L. and Katz, B., "The effect of sodium ions on the electrical activity of the giant axon of the squid", *J. Physiol.*, 1949, **108**, 37.

Hodgkin, A.L. and Keynes, R.D., "Active transport of cations in giant axons from Sepia and Loligo", *J. Physiol.*, 1955, **128**, 28.

Hofee, P., Englesberg, E. and Lamly, F., "The glucose permease system in bacteria", *Biochim. Biophys. Acta*, 1964, **79**, 337.

Hoffman, J.F., "The link between metabolism and the active transport of Na in human red cell ghosts", *Fed. Proc.*, 1960, **19**, 127.

Hoffman, J.F., "The active transport of sodium by ghosts of human red blood cells", *J. Gen. Physiol.*, 1962, **45**, 837.

Hoffman, J.F., "The red cell membrane and the transport of Na and K", *Am. J. Med.*, 1966, **41**, 666.

Hogben, C.A.M., "Observations on ionic movement through the gastric mucosa", *J. Gen. Physiol.*, 1968, **51**, 240, suppl.

Hokin, L.E. and Hokin, M.R., "Evidence for phosphatidic acid as the sodium carrier", *Nature*, 1959, **184**, 1068.

Hokin, L.E. and Hokin, M.R., "Diglyceride kinase and other pathways for phosphatidic acid synthesis in the erythrocyte membrane", *Biochim. Biophys. Acta*, 1963a, **67**, 470.

Hokin, L.E. and Hokin, M.R., Phosphatidic acid metabolism and active transport of sodium", *Fed. Proc.*, 1963b, **22**, 8.

Hokin, L.E. and Reosa, D., "Effect of preincubation of erythrocyte ghosts on ouabin-sensitive and ouabain-insensitive adenosine triphosphatase", *Biochim. Biophys. Acta*, 1964, **90**, 176.

Hong, S.K., Park, C.S., Park, Y.S. and Kim, J.K., "Seasonal changes of antidiuretic hormone action on sodium transport across frog skin", *Am. J. Physiol.*, 1968, **215**, 439.

Hoshiko, T , "Electrogenesis in frog skins", in *Biophysics of physiological and pharmacological actions* (Ed. Shanes, A.M.), Washington D.C., AAAS 1961.

Huf, E.G., "Versuche über den Zusammenhang zwischen Stoffwechsel, Potentialbildung und Funktion der Froschhaut", *Pflügers Arch. Ges. Physiol.*, 1935, **235**, 655.

Huf, E.G., Doss, N.S. and Wills, J.P., "Effects of metabolic inhibitors and drugs on ion transport and oxygen consumption in isolated frog skin", *J. Cell. Comp. Physiol.*, 1957, **56**, 43.

Hulcher, F.H., "Physical and chemical properties of myelin", *Arch. Biochem.*, 1963, **100**, 237.

Huxley, A. F. (1960), in A. K. Solomon (1960).

Huxley, A. F., "Excitation and conduction in nerve: quantitative analysis", *Science*, 1964, **145**, 1154.

Huang, C. and Thompson, T. E., "Properties of lipid bilayer membranes separating two aqueous phases: determination of membrane thickness", *J. Mol. Biol.*, 1965, **13**, 183.

Huang, C., Wheeldon, L. and Thompson, T. E., "The properties of lipid bilayer membranes separating two aqueous phases: formation of a membrane of simple composition", *J. Mol. Biol.*, 1964, **8**, 148.

Ilani, A., "Interaction between cations in hydrophobic solvent-saturated filters containing fixed negative charges", *Biophysical J.*, 1966, **6**, 329.

Jacobs, M. H., "Diffusion Processes", *Ergeb. Biol.*, 1935, **12**, 1.

Jacobson, B., "On the interpretation of dielectric constants of aqueous macromolecular solutions. Hydration of macromolecules", *J. Amer. Chem. Soc.*, 1955, **77**, 2919.

Jacquez, J. A., "The kinetics of carrier-mediated active transport of amino-acids", *Proc. Nat. Acad. Sci.*, 1961, **47**, 153.

Jacquez, J. A., "Carrier amino-acid stoichiometry in amino acid transport in Ehrlich ascites cells", *Biochim. Biophys. Acta*, 1963, **71**, 15.

Jacquez, J. A., "The kinetics of carrier-mediated transport: stationary-state approximations", *Biochim. Biophys. Acta*, 1964, **79**, 318.

Jakubovic, A. O., Hills, G. J., et Kitchener, J. A., "Coefficients d'auto-diffusion des ions dans les résines et les gels", *J. Chimie, Physique*, 1958, **55**, 263.

Jakubovic, A. O., Hills, G. J. and Kitchener, J. A., "Ionic mobilities in ion-exchange resins. Part 2: electrical conductivities of phenolsulphonic resins", *Trans. Faraday Soc.*, 1959, **55**, 1570.

Jardetzky, O., "On the definition of active transport", *Bull. Math. Biophys.*, 1960, **22**, 103.

Jardetzky, O. and Snell, F M., "Theoretical analysis of transport processes in living systems", *Proc. Nat. Acad. Sci.*, 1960, **46**, 616.

Jenny, H., "Kationen- und Anionenumtausch an Permutitgrenzflächen", *Kolloidchem. Beiheft*, 1927, **23**, 428.

Jenny, H., "Stodies on the mechanism of ion exchange in colloidal aluminum silicates", *J. Phys. Chem.*, 1932, **36**, 2217.

Jensen, W. A. and McLaren, A. D., "Uptake of proteins by plant cells, the possible occurrence of pinocytosis in plants", *Exptl. Cell. Res.*, 1960, **19**, 414.

Jorpes, E., "The protein component of the erythrocyte membrane or stroma", *Biochem. J.*, 1932, **26**, 1488.

Josefsson, J. O., "Some bioelectrical properties of amoeba proteus", *Acta Physiol. Scand.*, 1966, **66**, 395.

Kaneshiro, T. and Marr, A. G., "Phospholipids of Azotobacter agilis, Agrobacterium tumefaciens, and Escherichia coli", *J. Lipid. Res.*, 1962, **3**, 9.

Kanno, Y. and Loewenstein, W. R., "Low resistance coupling between gland cells. Some observations on intercellular contact membranes and intercellular space", *Nature*, 1964, **201**, 194.

Karnovsky, M. L., "Metabolic basis of phagocytic activity", *Physiol. Rev.*, 1962, **42**, 143.

Karreman, G. and Eisenman, G., "Electrical potentials and ionic fluxes in ion exchangers: I. "*n* type" non-ideal systems with zero current", *Bull. Math. Biophys.*, 1962, **24**, 413.

Katchalsky, A. and Curran, P. F., *Nonequilibrium Thermodynamics in Biophysics*, Harvard University Press, Cambridge, Massachusetts, 1965.

Katchalsky, A. and Kedem, O., "Thermodynamics of flow processes in biological systems", *Biophys. J.*, 1962, **2**, 53.

Katchalsky, A. and Spangler, R., "Dynamics of membrane processes", *Quart. Rev. Biophysics*, **1968**, **1**, 127.

Katz, B., *Nerve, muscle and synapse*, McGraw-Hill, New York, 1966.

Katzman, R. and Leiderman, P. H., "Brain potassium exchange in normal adult and immature rats", *Am. J. Physiol.*, 1953, **175**, 263.

Kavanau, J. L., *Structure and function in biological membranes*, Holden-Day, San Francisco, 1965.

Kaye, G. I., Pappas, G. E., Donn, A., and Mallett, N., "Studies on the cornea. II. The uptake and transport of colloidal particles by the living rabbit cornea in vitro", *J. cell. Biol.*, 1962, **12**, 481.

Kedem, O., "Criteria of active transport", *in Membrane transport and metabolism.* (Eds. Kleinzeller, A. and Kotyk, A.), London, Acad. Press, 1961.

Kedem, O. and Katchalsky, A., "Thermodynamic analysis of the permeability of biological membranes to nonelectrolytes", *Biochim. Biophys. Acta*, 1958, **27**, 229.

Kedem, O. and Katchalsky, A., "A physical interpretation of the phenomenological coefficients of membrane permeability", *J. Gen. Physiol.*, 1961, **45**, 143.

Kedem, O. and Katchalsky, A., "Permeability of composite membranes", *Trans. Faraday Soc.*, 1963a, **59**, 1918.

Kedem, O. and Katchalsky, A., "Permeability of composite membranes", *Trans. Faraday Soc.*, 1963b, **59**, 1931.

Kedem, O. and Katchalsky, A., "Permeability of composite membranes", *Trans. Faraday, Soc.*, 1963c, **59**, 1941.

Kelly, D.E., "Fine structure of desmosomes, hemidesmosomes and an adepidermal globular layer in developing newt epidermis", *J. Cell Biol.*, 1966, **28**, 51.

Kepes, A., "The place of permeases in cellular organization", in *The cellular functions of membrane transport.* (Ed. Hoffman, J.F.), Prentice-Hall, New Jersey, 1963.

Kepes, A. and Cohen, G.N., "Permeation", in *The bacteria* (Eds. Gunsalus, I.C. and Stanier, R.V.), Vol. 4. Academic Press, New York, 1962.

Keynes, R.D. and Lewis, P.R., "The resting exchange of radioactive potassium in crab nerve", *J. Physiol.*, 1951, **113**, 73.

Keynes, R.D. and Maisels, G.W., "The energy requirement for sodium extrusion from frog muscle", *Proc. Roy. Soc. London*, B, 1954, **142**, 383.

Khuri, R.N., "Glass microelectrodes and their uses in biological systems", in *Glass electrodes for hydrogen and other cations* (Ed. Eisenman, G.), Dekker, New York, 1967.

Kidder, G.W., Cereijido, M. and Curran, P.F., "Transient changes in electrical potentials differences across frog skin", *Am. J. Physiol.*, 1964, **207**, 935.

Kimizuka, H. and Koketsu, K., "Ion transport through cell membrane", *J. Theoret. Biol.*, 1964, **6**, 290.

Kinsolving, C.R., Post, R.L., and Beaver, D.L., "Sodium plus potassium transport adenosine triphosphatase activity in kidney", *J. Cell. Comp. Physiol.*, 1963, **62**, 85.

Kirkwood, J.G. and Oppenheim, I., *Chemical thermodynamics*, McGraw-Hill, New York, 1961.

Kirschner, L.B., "On the mechanism of active sodium transport across the frog skin", *J. Cell. Comp. Physiol.*, 1955, **45**, 61.

Klingenberg, M. and Schollmeyer, "ATP controlled redox states of respiratory carriers under the influence of DPNH-Hydrogen accepting substrates", *Biochim. Biophys. Res. Commun.*, 1961, **4**, 323.

Klotz, I.M., "Water", in *Horzons in biochemistry*, (Eds. Kasha, M. and Pullman, B.), Acad. Press, New York, 1962.

Koch, A.L., "The role of permease in transport", *Biochim. Biophys.* Acta, 1964, **79**, 117.

Koch, A. L., "Kinetics of permease catalyzed transport", *J. Theoret. Biol.*, 1967, **14**, 103.

Koefoed-Johnsen, V., "The effect of G-strophanthin (ouabain) on the active transport of sodium through the isolated frog skin", *Acta Physiol. Scand.*, 1958, **42**, 87 (Suppl. 145).

Koefoed-Johnsen, V. and Ussing, H. H., "The contributions of diffusion and flow to the passage of D_2O through living membranes. Effect of neuro-hypophyseal hormone on isolated anuran skin", *Acta Physiol. Scand.*, 1953, **28**, 60.

Koefoed-Johnsen, V. and Ussing, H. H., "Nature of the frog skin potential", *Acta Physiol. Scand.*, 1958, **42**, 298.

Kogl, F., de Gier, J., Mulder, J. I. and Van Deenen, L. M., "Metabolism and function of phosphatides. Specific fatty acid composition of the red blood cell membranes", *Biochem. Biophys. Acta*, 1960, **43**, 95.

Kolber, A. R. and Stein, W. D., "Identification of a component of a transport carrier system: isolation of the permease expression of the *lac* operon of Escherichia coli", *Nature*, 1966, **209**, 691.

Kono, T. and Colowick, S. P., "Isolation of skeletal muscle cell membrane and some of its properties", *Arch. Biochem. Biophys.*, 1961, **93**, 520.

Korn, E. D., "Structure of biological membranes", *Science*, 1966, **153**, 1491.

Korn, E. D. and Weisman, R. A., "Lipids and lipid metabolism. I. Loss of lipids during preparation of amoebae for electron microscopy", *Biochim. Biophys. Acta*, 1966, **116**, 309.

Kostyo, J. L. and Engel, F. L., "*In vitro* effects of growth hormone and carticotropin preparations on amino acid transport by isolated rat diaphragma", *Endocrinology*, 1960, **67**, 708.

Krishnamoorthy, C.,and Overstreet, R., "An experimental evaluation of ion exchange relationships", *Soil Science*, 1950, **69**, 41.

Kritchevsky, D. (Ed.), "Deuterium isotope effects in chemistry and biology", *Ann. N.Y. Acad. Sci.*, 1960, **84**, 573.

Krogh, A., "Osmotic regulation in the frog (R. esculenta) by active absorption of chloride ions", *Skand. Arch. Physiol.*, 1937, **76**, 60.

Krogh, A., "The active absorption of ions in some freshwater animals", *Z. Vergl. Physiol.*, 1938, **25**, 335.

Kushnir, L. D., "Studies on a material which induces electrical excitability in bimolecular membranes. I. Production, isolation, gross identification and assay", *Biochim. Biophys. Acta*, 1968, **150**, 285.

Lakshminarayanaiah, N., "Transport phenomena in artificial membranes", *Chem. Rev.*, 1965, **65**, 491.

Langeland, T., "On the application of irreversible thermodynamics to biological membrane phenomena", *Abst. Comm. Intern. Biophys. Congr. Estocolmo*, 1961.

Leaf, A., "Some actions of neurohypophysical hormones on a living membrane", *J. Gen. Physiol.*, 1960, **43**, 175, Suppl.

Leaf, A., Anderson, J. and Page, L. B., "Active sodium transport by the isolated toad bladder", *J. Gen. Physiol.*, 1958, **41**, 657.

Leaf, A., and Dempsey, E. F., "Some effects of mammalian neurohypophyseal hormones on metabolism and active transport of sodium by the isolated toad bladder", *J. Biol. Chem.*, 1960, **235**, 2160.

Leaf, A., and Hays, R. M., "Permeability of the isolated toad bladder to solutes and its modification by vasopressin", *J. Gen. Physiol.*, 1962, **45**, 921.

Leaf, A. and Renshaw, A., "Ion transport and respiration of isolated frog skin", *Biochem. J.*, 1957, **65**, 82.

Leb, D. E., Hoshiko, T., Lindley, B. D. and Dugan, J. A., "Effects of alkali metal cations on the potential across toad and bull frog urinary bladder", *J. Gen. Physiol.*, 1965, **48**, 527.

LeFevre, P. G., "Active transfer of glycerol and glucose across the human red cell membrane", *J. Gen. Physiol.*, 1948, **38**, 305.

LeFevre, P. G., "Persistence in erythrocyte ghosts of mediated sugar transport", *Nature*, 1961, **191**, 970.

LeFevre, P. G., "Upper limit for number of sugar transport sites in red cell surface", *Fed. Proc.*, 1961, **20**, 139.

LeFevre, P. G. and Davies, R. I., "Active transport into the human erythrocyte: evidence from comparative kinetics and competition among mono saccharides", *J. Gen. Physiol.*, 1951, **34**, 515.

LeFevre, P. G. and Marshall, J. K., "Conformational specificity in a biological sugar transport system", *Am. J. Physiol*, 1958, **194**, 333.

LeFevre, P. G. and McGinniss, G. F., "Tracer exchange vs net uptake of glucose through human red cell surface. New evidence for carrier mediated difusion", *J. Gen. Physiol.*, 1960, **44**, 87.

Lehninger, A. L., *The mitochondrion*, Benjamin, New York, 1964.

Lenard, J. and Singer, S. J., "Protein conformation in cell membrane preparations as studied by optical rotatory dispersion and circular dichroism", *Proc. Nat. Acad. Sci.*, 1966, **56**, 1828.

Lettvin, J. Y., Pickard, W. F., McCulloch, W. S. and Pitts, W., "A theory of passive ion flux through axon membranes", *Nature*, 1964, **222**, 1338.

Lev, A. A. and Buzhinski, E. P., "Cation specificity of model bimolecular phospholipid membranes with exposure to valinomycin", *Tsytologia (USSR)*, 1967, **9**, 102.

Levi, H. and Ussing, H. H., "The exchange of sodium and chloride ions across the fibre membrane of the isolated frog sartorius", *Acta Physiol. Scand.*, 1948, **16**, 232.

Levine, R., "Mechanisms of insulin action", *Diabetes*, 1961, **10**, 421.

Levinsky, N. and Sawyer, W., "Relation of metabolism of frog skin to cellular integrity and electrolyte transfer", *J. Gen. Physiol.*, 1953, **36**, 607.

Lewis, G. N., *Valence and the structure of atoms and molecules*, Chemical Catalog, Co., New York, 1923.

Lewis, G. N., and Randall, M., *Thermodynamics*, 2nd. ed., revised by K. S. Pitzer and L. Brewer. McGraw-Hill, New York, 1961.

Lewis, M. S. and Saroff, H. A., "The binding of ions to the muscle proteins. Measurements on the binding of potassium and sodium ions to myosin A, myosin B and actin", *J. Amer. Chem. Soc.*, 1957, **79**, 2112.

Lewis, W. H., "Pinocytosis", *Bull. J. Hopkins Hosp.*, 1931, **49**, 17.

Lidiard, A. B., "Ionic conductivity", *Handbuch der Physik*, Springer-Verlag, Berlin, 1957, Vol. 22.

Lindeman, B., "Sodium and calcium-dependence of threshold potential in frog skin excitation", *Biochim. Biophys. Acta*, 1968, **163**, 424.

Ling, G. N., "The interpretation of selective ion permeability and cellular potentials in terms of the fixed charge induction hypothesis", *J. Gen. Physiol.*, 1960, **43**, 1495.

Ling, G. N., *A physical theory of the living state*, Blaisdell, New York, 1962.

Ling, G. N., "Physiology and anatomy of the cell membrane: the physical state of water in the living cell", *Fed. Proc.*, 1965, **24**, S-103.

Ling, G. N., "Effects of temperature on the state of water in the living cell", in *Thermobiology* (Ed. A. Rose) Academic Press, New York, 1967.

Ling, G. N., "A new model for the living cell: a summary of the theory and recent experimental evidence in its support", *Intern. Rev. Cytology*, 1969a, **26** (in press).

Ling, G. N., "Measurements of potassium ion activity in the cytoplasm of living cells", *Nature*, 1969b, **221**, 386.

Ling, G. N. and Ochsenfeld, M. M., "Studies on the ionic permeability of muscle cells and their models", *Biophys. J.*, 1965, **5**, 777.

Ling, G. N. and Ochsenfeld, M. M., "Studies on ion accumulation in muscle cells", *J. Gen. Physiol.*, 1966, **49**, 819.

Ling, G. N., Ochsenfeld, M. M. and Karreman, G., "Is the cell membrane a universal rate-limiting barrier to the movement of water between the living cells and its surrounding medium?", *J. Gen. Physiol.*, 1967, **50**, 1807.

Loewenstein, W. R., "Permeability of membrane junctions", *Ann. N. Y. Acad. Sci.*, 1966, **137**, 441.

Loewenstein, W. R. and Kanno, Y., "Studies on an epithelial (gland) cell junction. I. Modifications of surface membrane permeability", *J. Cell. Biol.*, 1964, **22**, 565.

Loewenstein, W. R., Socolar, S. J., Higashino, S., Kanno, Y. and Davidson, N., "Intercellular communication: renal, urinary bladder, sensory, and salivary gland cells", *Science*, 1965, **149**, 295.

Luck, W., "The association of liquid water", *Fortschr. Chem. Forsch.*, 1964, **4**, 653.

Luzzati, V., "X-ray diffraction studies of lipid-water systems", in *Biological Membranes* (Ed. Chapman, D.) Academic Press, N. York, 1968.

Luzzati, V. and Husson, F., "The structure of liquid-crystalline phase of lipid-water systems", *J. Cell. Biol.*, 1962, **12**, 207.

Luzzati, V., Reiss-Husson, F., Rivas, E. and Gulik-Krzywicki, T., "Structure and polymorphism in lipid-water systems, and their possible biological implications", *Ann. N.Y. Acad. Sci.*, 1966, **137**, 409.

Luzzati, V. and Spegt, P. A , "Polymorphism of lipids", *Nature*, 1967, **215**, 701.

Luzzati, V., Tardieu, A., Gulik-Krzywicki, T., Rivas, E. and Reiss-Husson, F., "Structure of the cubic phases of lipid-water systems", *Nature*, 1968, **220**, 485.

Mackie, J. S. and Meares, P., "The diffusion of electrolytes in a cation-exchange resin membrane", *Proc. Roy. Soc. (A)*, 1955, **232**, 498.

MacRobbie, E. A. C., and Ussing, H. H., "Osmotic behaviour of the epithelial cells of frog skin", *Acta Physiol. Scand.*, 1961, **53**, 348.

Maddy, A. H., "The solubilization of the protein of the ox-erythrocyte ghost", *Biochim. Biophys. Acta*, 1964, **88**, 448.

Maddy, A. H., "The chemical organization of the plasma membrane of animal cells", in *Intern. Rev. of Cytol.* (Eds. Bourne, G. H. and Danielli, J. F.), Academic Press, New York, 1966a.

Maddy, A H., "The properties of the protein of the plasma membrane of ox eythrocytes", *Biochim. Biophys. Acta*, 1966b, **117**, 193.

Maddy, A. H., Huang, C. and Thompson, T. E., "Studies on lipid bilayer membranes: a model for the plasma membrane, *Fed. Proc.*, 1966, **25**, 933.

Malcolm, B. R., "Conformation of synthetic polypeptide and protein monolayers at interfaces", *Nature*, 1962, **195**, 901.

Marinetti, G. V and Gray, G. M., "A fluorescent chemical marker for the liver cell plasma membrane", *Biochim. Biophys. Acta*, 1967, **135**, 580.

Marro, F. and Pesente, L., "Rappresentazione del transporto attivo di sodio attraverso la cute isolata di rana in base alla termodinamica dei processi irreversibili. II. Deduzione di una relazione tra potenziale transcutaneo e corrente di corto-circuito", *Bollettino della Società Italiana di Biologia Sperimentale*, 1964a, **40**, 1443.

Marro, F. and Pesente, L., "Rappresentazione del transporto attivo di sodio attraverso la cute isolata di rana in base alla termodinamica dei processi irreversibili. I. Premesse e trattazione teorica", *Bolletino della Società Italiana di Biologia Sperimentale*, 1964b, **40**, 1440.

Marshall, J. M., "Intracellular transport in the amoeba Chaos Chaos", in *Intracellular transport* (Ed. Warren, K. B.), Academic Press, New York, 1966.

Mast, S. O. and Doyle, W. L., "Ingestion of fluid by amoeba", *Protoplasma*, 1934, **20**, 555.

Matsui, H. and Schwartz, A , "Partial purification of a highly active ouabain-sensitive Na^{+}–K^{+}-ATPase from cardiac muscle", *Fed. Proc.*, 1966, **25**, 622.

Maturana, H. R., "The fine anatomy of the optic nerve of Anurans: an electron microscope study", *J. Biophys. Biochem. Cytol.*, 1960, **7**, 107.

Mauro, A., "Space charge regions in fixed charge membranes and the associated property of capacitance", *Biophys. J.*, 1962, **2**, 179.

McBain, J. W. and Lee, W. W., "Vapor pressure data and phase diagrams and some concentrated soap-water systems above room temperature", *Oil and Soap*, 1943, **20**, 17.

McCollester, D. L. and Randle, P. J., "Isolation and some enzymic activities of muscle-cell membranes", *Biochem. J.*, 1961, **78**, 27P.

McConaghey, P. D. and Maizels, M., "Cation exchangers of lactose-treated human red cells", *J. Physiol.*, 1962, **162**, 485.

Meares, P., "Conductivity of a cation-exchange resin", *J. Polym. Sci.*, 1956, **20**, 507.

Mercer, E. H., "An electron microscopic study of amoeba proteus", *Proc. Roy. Soc. B*, 1959, **150**, 216.

Meyer, K.H. and Bernfeld, P., "The potentiometric analysis of membrane structure and its application to living animal membranes", *J. Gen. Physiol.*, 1946, **29**, 353.

Meyer, K.H. and Mark, H., *Makromolekulare Chemie*, 3d. ed., Akademische Verlagsanstalt, Leipzig, 1953.

Miller, D.M., "Sugar uptake as a function of cell volume in human erythrocytes", *J. Physiol.*, 1964, **170**, 219.

Miller, F., "Hemoglobin absorption by the cells of the proximal convoluted tubule in mouse kidney", *J. Biophysic. and Biochem. Cytol.*, 1960, **8**, 689.

Mitchel, P., "The chemical asymmetry of membrane transport precesses", in *Cell interface reactions.* (Ed. Brown H.D.), Scholar's Library, New York, 1963.

Miyamoto, V.K. and Thompson, T.E., "Some electrical properties of lipid bilayer membranes", *J. Colloid. Sci.*, 1967, **25**, 16.

Mohl (1846), quoted by Ling (1962).

Morel, F., "Interprétation de la mesure des flux d'ions à travers une membrane biologique comportant un "Compartiment" cellulaire; example des mouvements de sodium à travers la peau de grenouille", in *The Method of isotopic tracers applied to study of active ion transport*, Pergamon, London, 1958.

Morszynsky, J.R., Hoshiko, T. and Lindley, B.D., "Note on the Curie principle", *Biochim. Biophys. Acta*, 1963, **75**, 447.

Moskowitz, M. and Calvin, M., "Components and structure of the human red cell membrane", *Exptl. Cell. Res.*, 1952, **3**, 33.

Mueller, P. and Rudin, D.O., "Induced excitability in reconstituted cell membrane structure", *J. Theoret. Biol.*, 1963, **4**, 268.

Mueller, P. and Rudin, D.O., "Action Potential phenomena in experimental bimolecular lipid membranes", *Nature*, 1967a, **213**, 603.

Mueller, P., and Rudin, D.O., "Development of K^+–Na^+ discrimination in experimental bimolecular lipid membranes by macrocyclic antibiotics"; *Biochem. Biophys. Res. Common.*, 1967b, **26**, 398.

Mueller, P. and Rudin, D.O., "Resting and action potentials in experimental bimolecular lipid membranes", *J. Theor. Biol.*, 1968a, **18**, 222.

Mueller, P. and Rudin, D.O., "Action potentials induced in bimolecular lipid membranes", *Nature*, 1968b, **217**, 713.

Mueller, P., Rudin, D.O., Tien, H.T. and Wescott, W.C., Formation and properties of bimolecular lipid membranes. *Recent Progress in Surface Science*, Academic Press, New York, 1964.

Mueller, P., Rudin, D.O., Tien, H.T. and Wescott, W.C., "Reconstitution of excitable cell membrane structure *in vitro*", *Circulation*, 1962a, **26**, 1167.

Mueller, P., Rudin, D.O., Tien, H.T. and Wescott, W.C., Reconstitution of cell membrane structure *in vitro* and its transformation into an excitable system", *Nature*, 1962b, **194**, 979.

Muir, A.R. and Peters, A., "Quintuple-layered membrane junctions at terminal bars between endothelial cells", *J. Cell. Biol.*, 1962, **12**, 443.

Mullins, L.J., "The penetration of some cations into muscle", *J. Gen. Physiol.*, 1959, **42**, 817.

Mullins, L.J. and Awad, M.Z., "The control of the membranes potential of muscles fibres by the sodium pump", *J. Gen. Physiol.* 1965, **48**, 761.

Nägeli, C. und Cramer, C., *Pflanzenphysiologische Untersuchungen*, Schultess, Zurich, 1855.

Nageotte, J., "Lames élémentaires de la myéline en présence de l'eau", *Compt. Rend.*, 1927, **185**, 44.

Nakao, T., Nagano, K., Adachi, K. and Nakao, M., "Separation of two adenosine triphosphatases from erythrocyte membrane", *Biochim. Biophys. Res. Commun.*, 1963, **13**, 444.

Nakao, T., Tashima, Y., Nagano, K. and Nakao, M., "Highly specific sodium-potassium-activated adenosine triphosphatase from various tissues of rabbit", *Biochim. Biophys. Res. Commun.*, 1965, **19**, 755.

Nathans, D., Tapley, D.F., and Ross, J.E., "Intestinal transport of amino acids studied in vitro with $L-I^{131}$ monoiodotyrosine", *Biochim. Biophys. Acta*, 1960, **41**, 271.

Neihof, R. and Sollner, K. "A quantitative electrochemical theory of the electrolyte permeability of mosaic membranes composed of selectively anion-permeable and selectively cation-permeable parts and its experimental verification. I. An outline of the theory and its quantitative test in model systems with auxiliary electrodes", *J. Phys. Chem.*, 1950, **54**, 157.

Némethy, G. and Scheraga, H.A., "The structure of water and hydrophobic bonding in proteins. III. The thermodynamic properties of hydrophobic bonds in proteins", *J. Phys. Chem.*, 1962, **66**, 1773.

Némethy, G., Steinberg, I.Z. and Scheraga, H.A., "Influence of water structure and of hydrophobic interactions on the strength of side-chain H bonds in proteins", *Biopolymers*, 1963, **1**, 43.

Netter, H., "Über die Elektrolytgleichgewichte in elektiv ionenpermeablen Membranen und ihre biologische Bedeutung", *Pflügers Arch.*, 1928, **220**, 107.

Neurath, H. and Bull, H. B., "The surface activity of proteins", *Chem. Rev.*, 1938, **23**, 391.

Novikoff, A. B., "Mitochondria", in *The Cell* (Eds. Brachet, J. and Mirsky, A.), Vol. 2, Academic Press, New York, 1961.

O'Brien, J. S., "Cell Membranes: composition, structure, function", *J. Theoret. Biol.*, 1967, **15**, 307.

O'Brien, J. S. and Sampson, E. L., "Brain lipids. III. Lipid composition of the normal human brain; gray matter, white matter, and myelin", *J. Lipid. Res.*, 1965, **6**, 537.

Odor, D. L., "Uptake and transfer of particulate matter from the peritoneal cavity of the rat", *J. Biophys. Biochem. Cytol.*, 1956, **2**, 105.

Ohki, S. and Fukuda, N., "Interlayer-interaction for a lipid bilayer model", *J. Theoret. Biol.*, 1967, **15**, 362.

Onishi, T., "Extraction of actin- and myosin-like proteins from erythrocyte membrane", *J. Biochem. (Tokyo)*, 1962, **52**, 307.

Onsager, L., "Reciprocal relations in irreversible processes", *Phys. Rev.* 1931a, **37**, 405.

Onsager, L., "Reciprocal relations in irreversible processes II", *Phys. Rev.*, 1931b, **38**, 2265.

Opit, L. J. and Charnock, J. S., "A molecular model for a sodium pump", *Nature*, 1965, **208**, 471.

Orgel, L. E., "The hydrogen bond", in *Biophysical Science* (Ed. Oncley, J. L.), Wiley, New York, 1959.

Orloff, J. and Handler, J. S., "Mechanism of action of antidiuretic hormones on epithelial structures", in *The cellular functions of membrane transport* (Ed. Hoffman, J. F.), Prentice Hall, London, 1963.

Osterhaut, W. J. V., "Calculations of bioelectric potentials. V. Potentials in halicystis", *J. Gen. Physiol.*, 1939, **23**, 53.

Overton, E., "Über die allgemein osmotischen Eigenschaften der Zelle, ihre vermutlichen Ursachen und ihre Bedeutung für die Physiologie", *Vierteljahresschr. Naturforsch. Ges., Zurich*, 1899, **44**, 88.

Paganelli, C. V., and Solomon, A. K., "The rate of exchange of tritiated water across the human red cell membrane", *J. Gen. Physiol.*, 1957, **41**, 259.

Pagano, R. and Thompson, T. E., "Spherical lipid bilayer membranes", *Biochim. Biophys. Acta*, 1967, **144**, 666.

Page, E., "Cat heart muscle in vitro: II. The steady state resting potential in quiescent papillary muscles", *J. Gen. Physiol.*, 1962, **46**, 189.

Page, E., Goerke, R.J. and Storm, S.R., "Cat heart muscle in vitro. IV. Inhibition of transport in quiescent muscles", *J. Gen. Physiol.*, 1964, **47**, 531.

Page, E. and Storm, S., "Cat heart muscle in vitro. VIII. Active transport of sodium in papillary muscles", *J. Gen. Physiol.*, 1965, **48**, 957.

Palade, G.E., "Fine structure of blood capillaries", *J. Appl. Phys.*, 1953, **24**, 1424.

Palade, G.E., "The endoplasmic reticulum", *J. Biophys. Biochim. Cytol.*, 1956, **2**, 85.

Palay, S.L. and Karlin, L.J., "An electron microscopic study of the intestinal villus. I. The fasting animal", *J. Biophys. Biochem. Cytol.*, 1959, **5**, 363.

Palmer, K.J. and Schmitt, F.O., "X-ray diffraction studies on lipid emulsion", *J. cell. Comp. Physiol.*, 1941, **18**, 385.

Pappenheimer, J.R., "Passage of molecules through capillary walls", *Physiol. Rev.*, 1953, **33**, 387.

Pappenheimer, J.R., Renkin, E.M., and Borrero, L.M., "Filtration, diffusion and molecular sieving through peripheral capillary membranes", *Amer. J. Physiol.*, 1951, **167**, 13.

Park, C.R., Post, R.L., Kalhan, C.F., Wright, J.H., Johnson, L.H. and Morgan, H.E., "The transport of glucose and other sugars across cell membranes and the effect of insulin", *Ciba Found. Colloc. Endocrin.*, 1956, **9**, 240.

Parsegian, A., "Forces between lecithin bimolecular leaflets are due to a disordered surface layer", *Science*, 1967, **156**, 939.

Parsons, D.S. and Prichard, J.S., "Properties of some model systems for transcellular active transport", *Biochim. Biophys. Acta*, 1966, **126**, 471.

Passow, H., "Steady state diffusion of non-electrolytes through epithelial brush border", *J. Theoret. Biol.*, 1967, **17**, 383.

Passow, H., "The molecular basis of ionic selectivity of red cell membranes", *Proc. Symp. on The molecular basis of membrane function*, 1968.

Patlak, C.S., Goldstein, D.A. and Hoffman, J.F., "The flow of solute and solvent across a two-membrane system", *J. Theoret. Biol.*, 1963, **5**, 426.

Pauling, L., *The nature of the chemical bond*, Cornell Univ. Press, Ithaca, New York, 1960.

Pedersen, C.J., "Cyclic polyethers and their complexes with metal salts", *J. Am. Chem. Soc.*, 1967, **89**, 7017.

Penninston, J.T., Harris, R.A., Asai, J. and Green, D.E., "The conformational basis of energy conservation in membrane systems. The conformational changes in mitochondria", *Proc. Nat. Acad. Sci.*, 1968, **59**, 624.

Peters, A., "The structure of myelin sheaths in the central nervous system of Xenopus laevis (Baudin)", *J. Biophys. Biochem. Cytol.*, 1960, **7**, 121.

Peters, R., "Hormones and the cytoskelton", *Nature*, 1956, **177**, 426.

Pethica, B.A., "Surface activity and the microbial cell", *Soc. Chem. Ind., London, Monograph*, 1965, **19**, 85.

Pfeffer, W., *Osmotische Untersuchungen*, Engelmann, Leipzig, 1877.

Pfeffer, W., "Abhandl. mathematisch-physischen Klasse", *Königl. Sächs. Ges. Wis.* 1890, **16**, 185.

Pidot, A.L., and Diamond, J.M., "Streaming potentials in a biological membrane", *Nature*, 1964, **201**, 701.

Pimentel, G.C. and McClellan, A. L., *The hydrogen bond*, Reinhold, New York, 1960.

Podolsky, R.J., "The structure of water and electrolyte solutions", *Circulation*, 1960, **21**, 818.

Ponder, E., "Various types of ghosts derived from human red cells: heat fragmentation and phase optics studies", *J. Exptl. Biol.*, 1952, **29**, 605.

Post, R.L., Merrit, C.R., Kinsolving, C.R. and Albright, C.D., "Membrane adenosinetriphosphatase as a participant in the active transport of sodium and potassium in the human erythrocyte", *J. Biol. Chem.*, 1960, **235**, 1796.

Prigogine, I., *Introduction to Thermodynamics of Irreversible Processes*, Wiley, New York, 1961.

Purdom, L., Ambrose, E.J., and Klein, G., "A correlation between electrical surface charge and some biological characteristics during the stepwise progression of a mouse sarcoma", *Nature*, 1958, **181**, 1586.

Racker, E., "A mitochondrial factor conferring oligomycin sensitivity on soluble mitochondrial adenosintriphosphatase (ATPase)", *Biochem. Biophys. Res. Commun.*, 1963, **10**, 435.

Razin, J., Argaman, M. and Avigan, J., "Chemical composition of mycoplasma cells and membranes", *J. Gen. Microbiol.*, 1963, **33**, 477.

Reeves, R.E., "Cuprammonium-glycoside complexes", *Advan. Carbohydrate Chem.*, 1951, **6**, 107.

Rega, A.F., Garrahan, P.J. and Pouchan, M.I., "Effects of ATP and Na^+ on K^+ activated phosphatase from red blood cell membranes", *Biochim. Biophys. Acta*, 1968, **150**, 742.

Rehm, W. S., "Gastric potential and ion transport" in *Transcellular membrane potentials and ion fluxes*, (Ed. Snell, F. M.), Gordon and Breach, New York, 1964.

Reid, E. W., "Intestinal absorption of solutions", *J. Physiol.*, 1902, **28**, 241.

Reiser, A., *Hydrogen Bonding*, (Ed. Hadzi, D. and Thompson, H. W.) Pergamon Press, New York, 1959.

Reisin, I. and Cereijido, M., "Na fluxes in single, isolated seminal tubules of the rat", *Biophysical, J.*, 1969, **9**, A-164.

Reiss, E. and Kipnis, D., "The mechanism of action of growth hormone and hydrocortisone on protein synthesis in striated muscles", *J. Lab. Clin. Med.*, 1959, **54**, 937.

Reiss-Husson, F., "Structure des Phases liquide-cristallines de différents phospholipides monoglycérides, sphingolipides, anhydres ou en présence d'eau", *J. Molec. Biol.*, 1967, **25**, 363.

Rendi, R., and Uhr, M. L., "Sodium, potassium-requiring adenosinetriphosphatase activity", *Biochim. Biophys. Acta*, 1964, **89**, 520.

Renkin, E. M., "Filtration, diffusion and molecular sieving through porous cellulose membranes", *J. Gen. Physiol.*, 1954, **38**, 225.

Rickenberg, H. V., Cohen, G. N., Buttin, G. and Monod, J., "La galactoside-perméase d'Echerichia coli", *Ann. Inst. Pasteur*, 1956, **91**, 829.

Rickenberg, H. V., and Maio, J. J., "The transport of galactose by mammalian tissue culture cells", in *Membrane transport and metabolism.* (Eds. Kleinzeller, A. and Kotyk, A.) Academic Press, London, 1961.

Riggs, T. R., Walter, L. M. and Christensen, H. N., "Potassium migration and aminoacid transport", *J. Biol. Chem.*, 1958, **233**, 1479.

Riklis, E. and Quastel, J. H., "Effect of cations on sugar absorption by isolated surviving guinea pig intestine", *Can. J. Biochem. Physiol.*, 1958, **36**, 347.

Robertson, J. D., "The molecular biology of cell membranes", in *Molecular Biology* (Ed. Nachmansohn, D.), Academic Press, New York, 1960.

Robertson, J. D., "The membrane of the living cell", *Scient. American*, 1962, **206**, 64 (April).

Robertson, J. D., "Unit membranes: a review with recent new studies of experimental alterations and a new subunit structure in synaptic membranes", in *Cellular Membranes in Development* (Ed. Locke, M.), Academic Press, New York, 1964.

Roelofsen, B., Baadenhuysen, H., and van Deenen, L. L. M., "Effects of organic solvents on the adenosine triphosphatase activity of erythrocyte ghosts", *Nature*, 1966, **212**, 1379.

Rojas, E. and Tobias, J.M., "Membrane model: association of inorganic cations with phospholipid monolayers", *Biochim. Biophys. Acta*, 1965, **94**, 394.

Rosenberg, T., "The concept and definition of active transport", *Symposia Soc. Exptl. Biol.*, 1954, **8**, 27.

Rosenberg, T.H. and Wilbrandt, W., "The kinetics of membrane transports involving chemical reactions", *Exper. Cell. Res.*, 1955, **9**, 49.

Rosenberg, T., Wilbrandt, W., "Uphill transport induced by counterflow", *J. Gen. Physiol.*, 1957, **41**, 289.

Rosenberg, T. and Wilbrandt, W., "Carrier transport uphill", *J. Theoret. Biol.*, 1963, **5**, 288.

Rothmund, V. and Kornfeld, G., "Der Basenaustausch im Permutit", *I.Z. Anorg. Allgem. Chem.*, 1918, **103**, 129.

Rothstein, A., "Enzyme system of the cell surface inolved in the uptake of sugars by yeast", *Symp. Soc. Exp. Biol.*, 1954, **8**, 165.

Rotunno, C.A., Pouchan, M.I. and Cereijido, M., "Location of the mechanism of active transport of sodium across the frog skin", *Nature*, 1966, **210**, 597.

Rotunno, C.A., Kowalewski, V. and Cereijido, M., "Nuclear spin resonance evidence for complexing of sodium in frog skin", *Biochim. Biophys. Acta*, 1967, **135**, 170.

Rustad, R.C., "The physiology of pinocytosis", in *Recent progress in surface science*, Vol.2. (Ed. Danielli, J.F.) Academic Press, New York, 1964.

Salem, L., "The role of long-range forces in the cohesion of lipoproteins. *In* Symposium on the nature of lipoproteins", *Can. J. Biochem. Physiol.*, 1962, **40**, 1287.

Salton, M.R.J. and Freer, J.H., "Composition of the membranes isolated from several gram-positive bacteria", *Biochim. Biophys. Acta*, 1965, **107**, 531.

Sbarra, A.J., Maney, B., and Shirley, W., "The metabolism of leukemic cells during phagocytosis", *Bact. Proc.*, 1961, **61**, 137.

Sbarra, A.J., Shirley, W. and Bardawil, W.A., "Piggy-back' phagocytosis", *Nature*, 1962, **194**, 255.

Schachtschabel, P., "Adsorption by clay minerals", *Kolloidchemische Beihefte*, 1940, **51**, 199.

Schatzman, J.H., "Herzglykoside als Hemmstoffe für den aktiven Kalium- und Natriumtransport durch die Erythrocytenmembran", *Helv. Physiol. Acta*, 1953, **11**, 346.

Scheer, B.T., "The flux-force relations across complex membranes with active transport", *Bull. Math. Biol.*, 1960, **22**, 269.

Schlögl, R., "The theory of the diffusion potentials and ion transport in free solution and in charged membranes", *Z. Elektrochem.*, 1954, **58**, 672.

Schlögl, R., "The significance of convection in transport processes across porous membranes", *Disc. Faraday Soc.*, 1956, **21**, 46.

Schlögl, R., "Stofftransport durch Membranen", *Forts. Phys. Chem.*, 1964, **9**, I.

Schmidt, W.J., "Doppelbrechung und Feinbau der Markscheide der Nervenfasern", *Z. Zellforsch.*, 1936, **23**, 657.

Schmitt, F.O., "Molecular organization of the nerve fiber", in *Biophysical Science. A study program* (Ed. Oncley, J.L.), John Wiley, New York, 1959.

Schmitt, F.O. and Bear, R.S., "The optical properties of vertebrate nerve axons as related to fiber size", *J. Cellular Comp. Physiol.*, 1937, **9**, 261.

Schmitt, F.O., Bear, R.S. and Palmer, K.J., "X-ray diffraction studies on the structure of the nerve myelin sheath", *J. Cell. Comp. Physiol.*, 1941, **18**, 3.

Schmitt, F.O., Bear, R.S. and Ponder, E., "Optical properties of the red cell membrane", *J. Cell. Comp. Physiol.*, 1936, **9**, 89.

Schneider, L. and Wohlfarth-Bottermann, K.E., "Proteinstudien. IX. Elektronenmikroskopische Untersuchungen in Amöben unter besonderer Berücksichtigung der Feinstruktur des Cytoplasmas", *Protoplasma*, 1959, **51**, 377.

Schulman, J.H. and Hughes, A.H., "Monolayers of proteolitic enzymes and proteins. III. Enzyme reactions and penetration of protein monolayers", *Biochem. J.*, 1935, **29**, 1236.

Schulman, J.H. and Rideal, E.K., "Monomolecular interaction in monolayers, I. Complexes between large molecules", *Proc. Roy. Soc. B.* 1937, **122**, 29.

Schultz, S.G., Curran, P.F., Chez, R.A. and Fuisz, R.E., "Alanine and sodium fluxes across mucosal border of rabbit ileum", *J. Gen. Physiol.*, 1967, **50**, 1241.

Schultz, S.G., Epstein, W. and Solomon, A K., "Cation transport in Escherichia coli. IV. Kinetics of net K uptake", *J. Gen. Physiol.*, 1963, **47**, 329.

Schultz, S. and Zalusky, R., "The interaction between active sodium transport and active sugar transport in the isolated rabbit ileum", *Biochim. Biophys. Acta*, 1963, **71**, 503.

Schumaker, V. N., "Uptake of protein from solution by the amoeba proteus", *Exptl. Cell. Res.*, 1958, **15**, 314.

Schwartz, A., "A sodium and potassium-stimulated adenosine triphosphatase from cardiac tissues. I. Preparation and properties", *Biochim. Biophys. Res. Commun.*, 1962, **9**, 301.

Schwartz, I. L. and Walter, R., "Factors influencing the reactivity of the toad bladder to the hydro-osmotic action of vasopressin", *Am. J. Med.*, 1967, **42**, 769.

Schwartz, T. L. and Snell, F. M., "Non-steady-state three compartment tracer kinetics", *Biophysical J.*, 1968, **8**, 805.

Sen, A. K. and Post, R. L., "Stoichiometry of active Na^+ and K^+ transport to energy-rich phosphate break down in human erythrocytes", *Fed. Proc.*, 1961, **20**, 138.

Sen, A. K. and Widdas, W. F., "Variations of the parameters of glucose transfer across the human erythrocyte membrane in the presence of inhibitors of transfer", *J. Physiol.*, 1962, **140**, 404.

Sessa, G. and Weissmann, G., "Effect of polyene antibiotics on phospholipid spherules containing varying amounts of charged components", *Biochim. Biophys. Acta*, 1966, **135**, 416.

Shah, D. O. and Schulman, J. H., "Binding of metal ions to monolayers of lecithins plasmalogen, cardiolipin, and dicetyl phosphate", *J. Lipid. Res.*, 1965, **6**, 341.

Shah, D. O. and Schulman, J. H., "The ionic structure of lecithin monolayers", *J. Lipid. Res.*, 1967, **8**, 227.

Shanes, A. M. and Bianchi, C. P., "The distribution and kinetics of release of radiocalcium in tendon and skeletal muscle". *J. Gen. Physiol.*, 1959, **42**, 1123.

Shaw, T. I., *Ph. D. Thesis*, Cambridge University, 1954.

Shea, S. M. and Karnovsky, M. J., "Brownian motion: a theoretical explanation for the movement of vesicles across the endothelium", *Nature*, 1966, **212**, 353.

Shean, G. M. and Sollner, K., "Carrier mechanisms in the movement of ions across porous and liquid ion exchanger membranes", *Ann. N. York Acad. Sci.*, 1966, **137**, 759.

Shemyakin, M. M., Ovchinnikov, Y. A., Ivanov, V. T., Antonov, V. K., Shkrob, A. M., Mikhaleva, I. I., Estratov, A. V. and Malenkov, G. G., "The physicochemical basis of the functioning of biological membranes: conformational specificity of the interaction of cyclodepsipeptides with membranes and of their complexation with alkali metal ions", *Biochim. Biophys. Research Communications*, 1967, **29**, 834.

Sheppard, C.W., "The theory of the study of transfers within a multicompartment system using isotopic tracers", *J. Appl. Phys.*, 1948, **19**, 70.

Shewman, P.G., *Diffusion in solids*, McGraw-Hill, New York, 1963.

Sidel, V.W. and Solomon, A.K., "Entrance of water into human red cells under an osmotic pressure gradient", *J. Gen. Physiol.*, 1957, **41**, 243.

Silverman, M. and Goresky, C.A., "A unified kinetic hypothesis of carrier mediated transport: its applications", *Biophysical J.* 1965, **4**, 487.

Sjodin, R.A., "Rubidium and cesium fluxes in muscle as related to the membrane potential", *J. Gen. Physiol.*, 1959, **42**, 983.

Sjodin, R.A., "Some cations interactions in muscle", *J. Gen. Physiol.*, 1961, **44**, 929.

Sjodin, R.A., "The potassium flux ratio in skeletal muscle as a test for independent ion movement", *J. Gen. Physiol.*, 1965, **48**, 777.

Sjöstrand, F.S., "Ultrastructure and function of cellular membranes", in *The Membranes* (Ed. Dalton, A.J. and Hagenau, F.), Academic Press, New York, 1968.

Skou, J.C., "The influence of some cations on the adenosine triphosphatase from peripheral nerves", *Biochim. Biophys. Acta*, 1957, **23**, 394.

Skou, J.C., "Further investigations on a (Mg^{++} + Na^{+}) activated adenosine triphosphatase possible related to the active linked transport of Na^{+} and K^{+} across the nerve membrane", *Biochim. Biophys. Acta*, 1960, **42**, 6.

Skou, J.C., "Preparation from mammalian brain and kidney of the enzyme system involved in active transport of Na^{+} and K^{+}", *Biochim. Biophys. Acta*, 1962, **58**, 314.

Skou, J.C., "Studies on the Na^{+} + K^{+} activated ATP hydrolyzing enzyme system. The role of SH groups", *Biochem. Biophys. Res. Commun.*, 1963a, **10**, 79.

Skou, J.C., "Enzymatic aspects of active transport of Na^{+} and K^{+} across the cell membrane", in *Drugs and Membranes*, (Ed. Hogben, C.A.M.), Ist Intern. Congr. Pharmacol. Stockholm, 1961, Pergamon Press, London, 1963b, Vol.4, p.41.

Skou, J.C., "Enzymatic aspects of active linked transport of Na^{+} and K^{+} across the cell membrane", *Symposium on permeability*, Wageninge, 1963c.

Skou, J.C., "Enzymatic basis for active transport of Na^{+} and K^{+} across cell membrane", *Physiol. Rev.*, 1965, **45**, 596.

Skoulios, A.E. and Luzzati, V., "La structure des colloides d'association. III. Descriptive des phases mesomorphes des savons de sodium purs, recontres au-dessus de 100°C", *Acta Crystallogr.*, 1961, **14**, 278.

Slautterback, D.B., "Mitochondria in cardiac muscle cells of the canary and some other birds", *J. Cell. Biol.*, 1965, **24**, 1.

Small, D.M., "A classification of biologic lipids based upon their interaction in aqueous systems", *J. Am. Oil. Chem. Soc.*, 1967, **45**, 108.

Snell, F.M. and Leeman, C.P., "Temperature coefficients of the sodium transport system of isolated frog skin", *Biochim. Biophys. Acta*, 1957, **25**, 311.

Sollner, K.S., Dray, S., Grim, E. and Neihof, R., in *Ion transport across membranes*, (ed. Clarke, H.T.), Academic Press, New York, 1954.

Solomon, A.K., "Compartmental methods of kinetical analysis", in *Mineral metabolism* (Ed. Comar, C.L. and Bronner, F.), Academic Press, New York, 1960).

Solomon, A.K., "Ion transport in single population", *Biophysical J.*, 1962, **2**, 79.

Solomon, A.K., "Validity of tracer measurements of fluxes in kidney tubules and other three compartments systems", in *Transcellular membrane potentials and ionic fluxes* (Ed. Snell, F.M. and Noell, W.K.), Gordon and Breach, New York, 1964).

Solomon, A.K., "Characterization of biological membranes by equivalent pores", *J. Gen. Physiol.*, 1968, **51**, 335, suppl.

Spangler, R.A. and Snell, F.M., "Transfer function analysis of an oscillatory model chemical system", *J. Theor. Biol.*, 1967, **16**, 381.

Standish, M.M. and Pethica, B.A., "Surface pressure and surface potential study of a synthetic phospholipid at the air/water interface", *Trans. Faraday Soc.*, 1968, **64**, 1113.

Staverman, A.J., "The theory of measurement of osmotic pressure" *Rec. Trav. Chim.*, 1951, **70**, 344.

Stein, W.D., "The permeability of erythrocyte ghosts", *Exptl. Cell. Res.*, 1956, **11**, 232.

Stein, W.D., "Diffusion and Osmosis", *in Comprehensive Biochemistry*. (Ed. Florkin, M. and Stotz, E.H.) Vol 2. Chapter 3. Elsevier, Amsterdam, 1962a.

Stein, W.D., "Spontaneous and enzyme induced dimer formation and its role in membrane permeability. III. The mechanism of movement of glucose across the human erythrocyte membrane", *Biochim. Biophys. Acta*, 1962b, **59**, 66.

Stein, W.D., *The movement of molecules across cell membranes*, Academic Press, New York, 1967.

Stein, W.D., and Danielli, J.F., "Structure and function in red cell permeability", *Disc. Faraday Soc.*, 1956, **21**, 238.

Stoeckenius, W., "Structure of the plasma membrane, an electron microscope study", *Circulation*, 1962a, **26**, 1066.

Stoeckenius, W., "The molecular structure of lipid-water systems and cell membrane models, studied with the electron microscope", in *The interpretation of ultrastructure*. (Ed. Harris, J.C.) Academic Press, New York, 1962b.

Sutherland, E.W., and Rall, T.W., "The relation of adenosine-3′-5′-phosphate and phosphorylase to the actions of catecholamines and other hormones", *Pharmacol. Rev.*, 1960, **12**, 265.

Szabo, G., Eisenman, G., and Ciani, S.M., "Ion distribution equilibria in bulk phases and the ion transport properties of bilayer membranes produced by neutral macrocyclic antibiotics", *Proc. Coral Gables Conf. on the Principles of Biological Membranes*, Gordon and Breach, N.York, 1969 in press.

Taft, R.W., "The general nature of the proportionality of polar effects of substituent groups in organic chemistry", *J. Amer. Chem. Soc.*, 1953, **75**, 4231.

Tasaki, I. and Singer, I., "Membrane macromolecules and nerve excitability: a physico-chemical interpretation of excitation in squid giant axons", *An. N.Y. Acad. Sci.*, 1966, **137**, 792.

Tasaki, I., and Takenaka, T., "Ion fluxes and excitability in squid giant axon", in *The Cellular functions of Membrane transport* (Ed. Hoffman, J.F.) Prentice Hall, Englewood Cliffs, 1964.

Tasaki, I., Teorell, T. and Spyropoulos, C.S., "Movement of radioactive tracers across a squid axon membrane", *Am. J. Physiol.*, 1961, **200**, 11.

Tasaki, I., Watanabe, A., Sandlin, R. and Carnay, L., "Changes in fluorescence turbidity, and birefringence associated with nerve excitation", *Proc. Nat. Acad. Sci.*, 1968, **61**, 883.

Taube, H., "Steric problems in the hydration of ions in solution", *Prog. Stereochem.*, 1962, **3**, 95.

Teorell, T., "A method of studying conditions within diffusion layers", *J. Biol. Chem.*, 1936, **113**, 735.

Teorell, T., "Permeability properties of erythrocyte ghosts", *J. Gen. Physiol.*, 1952, **35**, 669.

Teorell, T., "Transport processes and electrical phenomena in ionic membranes", *Progress in Biophysics and Biophysical Chemistry*, 1953, **3**, 305.

Teorell, T., "A contribution to the knowledge of rhythmical transport processes of water and salts", *Exp. Cell. Res. (Suppl.)* 1955, **3**, 339.

Teorell, T., "Transport phenomena in membranes", *Disc. Faraday Soc.*, 1956, **21**, 22.

Teorell, T., "Transport processes in membranes in relation to the nerve mechanism", *Exptl. Cell. Res.*, 1958, **5**, 83 Suppl. 5.

Teorell, T., "Electrokinetic membrane processes in relation to properties of excitable tissues. I. Experiments on oscillatory transport phenomena in artificial membranes", *J. Gen. Physiol.*, 1959, **42**, 831.

Teorell, T., "Excitability phenomena in artificial membranes", *Biophys. J.*, 1962, **2**, 27 (part 2).

Thompson, J. F., Biological effects of Deuterium, Macmillan, New York, 1966.

Tien, H. T. and Diana, A. L., "Black lipid membrane in aqueous media: the effect of salts on electrical properties", *J. Colloid Sci.*, 1967, **24**, 287.

Tosteson, D. C., Andreoli, T. E., Tieffenberg, M. and Cook, P., "The effects of macrocyclic compounds on cation transport in sheep red cells and thin and thick lipid membranes", *J. Gen. Physiol.*, 1968, **51**, 373 Suppl.

Troshin, A. S., *Problems of cell permeability*, Pergamon, London, 1966.

Ungar, G., Ascheim, E., Psychoyos, S. and Romano, D. V., "Reversible changes in protein configuration in stimulated nerve structure", *J. Gen. Physiol.*, 1957, **40**, 635.

Ungar, G. and Romano, D. V., "Fluorescence changes in nerve induced by stimulation; their relation to protein structure", *J. Gen. Physiol.*, 1962, **46**, 267.

Ussing, H. H., "The use of tracers in the study of active ion transport across animal membranes", *Cold Spr. Harb. Symp. Quant. Biol.*, 1948, **13**, 193.

Ussing, H. H., "The active transport through the isolated frog skin in the light of tracer studies", *Acta Physiol. Scand.*, 1949a, **17**, 1.

Ussing, H. H., "Distinction by means of tracers between active transport and diffusion", *Acta Physiol. Scand.*, 1949b, **19**, 43.

Ussing, H. H , "Active transport of inorganic ions", *Symp. Soc. Exper. Biol.*, 1954, **8**, 407.

Ussing, H. H., "Active and passive transport across epithelial membranes", in *The method of isotopic tracers applied to the study of active ion transport.* (Ed. Coursaget, J.) Pergamon, London, 1958.

Ussing, H.H., and Windhager, E.E., "Nature of shunt path and active sodium transport path through frog skin epithelium", *Acta physiol. Scand.*, 1964, **61**, 484.

Ussing, H.H. and Zerahn, K., "Active transport of sodium as the source of electric current in the short-circuited isolated frog-skin", *Acta Physiol. Scand.*, 1951, **23**, 110.

Vandenheuvel, F.A., "Study of biological structure at the molecular level with stereomodel projections. I. The lipids in the myelin sheath of nerve", *J. Amer. Oil Chem. Soc.*, 1963, **40**, 455.

Vandenheuvel, F.A., "Structural studies of biological membranes. The structure of myelin", *Ann. N. Y. Acad. Sci.*, 1965, **122**, 57.

Vargas, F.F., "Water flux and electrokinetic phenomena in the squid axon", *J. Gen. Physiol.*, 1968, **51**, 123, Suppl.

Vargas, F.F. and Johnson, J.A., "An estimate of reflection coefficients for rabbit heart capillaries", *J. Gen. Physiol.*, 1964, **47**, 667.

Verwey, E.J.W. and Overbeek, J.T.G., *Theory of the stability of lyophobic colloids*, Elsevier, Amsterdam, 1948.

Vidaver, G.A., "Glycine transported by hemolyzed and restored pigeon red cells", *Biochemistry*, 1964, **3**, 795.

Vilallonga, F., "Surface chemistry of L-α-dipalmitoyl lecithin at the air-water interface", *Biochim. Biophys. Acta*, 1968, **163**, 290.

Vilallonga, F., Altschul, R. and Fernández, M., "Lipid-protein interaction at the air-water interface", *Biochim. Biophys. Acta*, 1967, **135**, 406.

Vilallonga, F., Fernández, M., Rotunno, C.A. and Cereijido, M., "Interactions of L-α-dipalmitoyl lecithin monolayers with Na^+, K^+ or Li^+ and its possible role in membrane phenomena", *Biochim. Biophys. Acta*, 1969, **183**, 90.

Villegas, R. and Barnola, F.V., "Equivalent pore radius in the axolemma of the giant axon of the squid", *Nature*, 1960, **188**, 762.

Villegas, R., and Barnola, F.V., "Characterization of the resting axolemma in the giant axon of the squid", *J. Gen. Physiol.*, 1961, **44**, 963.

Villegas, R., Bruzual, I.B. and Villegas, G., "Equivalent pore radius of the axolemma of resting and stimulated squid axons", *J. Gen. Physiol.*, 1968, **51**, 81, Suppl.

Villegas, R., Caputo, C. and Villegas, L., "Diffusion barriers in the squid nerve fiber", *J. Gen. Physiol.*, 1962, **46**, 245.

Villegas, L. and Sananes, L., "Ionic permeability and water movement in frog gastric mucosa", *J. Gen. Physiol.*, 1968, **51**, 226, Suppl.

Walker, J.L. and Eisenman, G., "A test of the theory of the steady-state properties of an ion exchange membrane with mobile sites and dissociated counterions", *Biophys. J.*, 1966, **6**, 513.

Walker, B.L. and Kummerow, F.A., "Erythrocyte fatty acid composition and apparent permeability to non-electrolytes", *Proc. Soc. Exptl. Biol. Med.*, 1964, **115**, 1099.

Wallach, D.F.H., "Isolation of plasma membranes of animal cells", in *The Specificity of Cell Surfaces* (Ed. Davies, B.D. and Warren, L.), Prentice Hall, New Jersey, 1965.

Wallach, D.F.H. and Eylar, E.H., "Sialic acid in the cellular membranes of Ehrlich ascites-carcinoma cells", *Biochim. Biophys. Acta*, 1961, **52**, 594.

Wallach, D.F.H. and Kamat, V.B., "Plasma and cytoplasmic membrane fragments from Ehrlich ascites carcinoma", *Proc. Nat. Acad. Sci.*, 1964, **52**, 721.

Wallach, D.F.H. and Zahler, P.H., "Protein conformations in cellular membranes", *Proc. Nat. Acad. Sci.*, 1966, **56**, 1552.

Waug, D.F. and Schmitt, F.O., "Investigation of the thickness and ultrastructure of cellular membrane by the analytical leptoscope", *Cold Spring Harbor Symposia Quant. Biol.*, 1940, **8**, 17.

Ways, P. and Hanahan, D.J., "Characterization and quantification of red cell lipids in normal man", *J. Lipid. Res.*, 1964, **5**, 318.

Weidmann, S., *Elektrophysiologie der Herzmuskelfaser*, Medizinischer Verlag Hans Huber Bern, 1956.

Weiss, P., "Cell Contact", *Internat. Rev. Cytol.*, 1958, **7**, 391.

Wellings, S.R. and Deome, K.B., "Milk protein droplet formation in the Golgi apparatus of the C3H/Crgl Mouse mammary epithelial cells", *J. Biophys. Biochem. Cytol.*, 1961, **9**, 479.

Whittam, R., and Ager, M.E., "The connexion between active cation transport and metabolism in erythrocytes", *Biochem. J.*, 1965, **97**, 214.

Whittam, R., Edwards, B.A. and Wheeler, K.P., "An approach to the study of enzyme action in artificial membranes", *Biochem. J.* 1968, **107**, 3P.

Whittembury, G., "Action of antidiuretic hormone on the equivalent pore radius at both surfaces on the epithelium of the isolated toad skin", *J. Gen. Physiol.*, 1962, **46**, 117.

Whittembury, G., "Electrical potential profile of the toad skin epithelium", *J. Gen. Physiol.*, 1964, **47**, 795.

Whittembury, G., "Sodium and water transport in kidney proximal tubular cells", *J. Gen. Physiol.*, 1968, **51**, 303, Suppl.

Widdas, W. F., "Inability of diffusion to account for placental glucose transfer in the sheep and consideration of the kinetics of a possible carrier transfer", *J. Physiol.*, 1952, **118**, 23.

Widdas, W. F., "Facilitated transfer of hexoses across the human erythrocyte membrane", *J. Physiol.*, 1954, **125**, 163.

Widdas, W. F., "Hexose permeability of foetal erythrocytes", *J. Physiol.*, 1955, **127**, 318.

Wiegner, G. and Jenny, H., "Über Basenaustausch an Permutiten", *Kolloid, Z.*, 1927, **42**, 268.

Wiener, J., Spiro, D. and Loewenstein, W. R., "Studies on an epithelial (gland) cell junction. II. Surface structure", *J. Cell. Biol.*, 1964, **22**, 587.

Wilbrandt, W., "The significance of the structure of a membrane for its selective permeability", *J. Gen. Physiol.*, 1935, **18**, 933.

Wilbrandt, W., "Permeabilitätsprobleme", *Arch. Exp. Path. Pharmacol.*, 1950, **212**, 9.

Wilbrandt, W., "Secretion and transport of non-electrolytes", *Symp. Soc. Exp. Biol.*, 1954, **8**, 136.

Wilbrandt, W., "Transport of sugars across cellular and biological membranes", in *Membrane transport and Metabolism.* (Eds. Kleinzeller, A., Kotyk, A.) London, Academic Press, 1961.

Wilbrandt, W. and Rosenberg, T., "Weitere Untersuchungen über die Glucosepenetration durch die Erythrocytenmembran", *Helv. Physiol. Acta*, 1950, **8**, 82.

Wiseman, G., "Active stereochemically selective absorption of amino-acids from rat small intestine", *J. Physiol.*, 1951, **114**, 7P.

Wolfe, L. S., "Cell membrane constituents concerned with transport mechanisms", *Can. J. Biochem. Physiol.*, 1964, **42**, 971.

Wolff, J., "Transport of iodide and others anions in the thyroid gland", *Physiol. Rev.*, 1964, **44**, 45.

Wong, J. T., "The possible role of polyvalent carriers in cellular transports", *Biochim. Biophys. Acta*, 1965, **94**, 102.

Yos, J. M., Bade, W. L. and Jehle, H., "Specificity of the London-Eisenschitz-Wang forces". *Proc. Nat. Acad. Sci.*, 1957, **43**, 341.

Yost, W., *Diffusion in solids, liquids, gases*, Academic Press, New York, 1952.

Zadunaisky, J. A., "Active transport of chloride in frog cornea", *Am. J. Physiol.*, 1966, **201**, 506.

Zadunaisky, J. A. and Candia, O. A., "Active transport of sodium and chloride by the isolated skin of the South American frog Leptodactylus ocellatus, *Nature*, 1962, **195**, 1004.

Zadunaisky, J. A., Candia, O. A. and Chiarandini, D. J., "The origin of the short-circuit current in the isolated skin of the South American frog Leptodactylus ocellatus", *J. Gen. Physiol.*, 1963, **47**, 393.

Zadunaisky, J. A., Parisi, M. N. and Montoreano, R., "Effect of antidiuretic hormone on permeability of single muscle fibres", *Nature*, 1963, **200**, 365.

Zerahn, K., "Oxigen consumption and active sodium transport in the isolated and short circuited frog skin", *Acta Physiol. Scand.*, 1956, **36**, 300.

Zwolinski, B. J., Eyring, H. and Reese, C. E., "Diffusion and Membrane permeability, I", *J. Phys. Colloid. Chem.*, 1949, **53**, 1426.